Contemporary Electronics and Electrical Engineering

General Editors:
Professor A. H. W. Beck
Professor J. Lamb

Charge-Transfer Devices

Charge-Transfer Devices

G. S. Hobson B.A., Ph.D.
Reader,
Department of Electronic and Electrical Engineering,
University of Sheffield

A HALSTED PRESS BOOK

JOHN WILEY & SONS
New York

Published in the U.S.A.
by Halsted Press, a Division of
John Wiley & Sons, Inc., New York

First published in 1978 by
Edward Arnold (Publishers) Ltd
London

Library of Congress Cataloging in Publication Data

Hobson, Geoffrey Stanley
Charge-transfer devices. – (Contemporary electronics and electrical engineering).
1. Charge transfer devices (Electronics)
I. Title II. Series
621.381'73 TK7871.99.C45 78–40587

ISBN 0 470-26458-6

Printed in Great Britain

Preface

Charge-transfer devices, a development of integrated circuit engineering, provide the electronic system designer with a practical and economic means of simultaneously comparing features of analogue signals at many different times. As with other applications of integrated circuits, a close integration has been forced on to the semiconductor device technologist, the circuit designer and the system designer. There is such a strong interaction between these three operations, in combining useful system function with technological capability, that they are virtually inseparable.

The objective of this book is to examine the physical behaviour of the components of charge-transfer device engineering by using simple models, so avoiding the inevitable detail of quantitatively accurate descriptions. It is assumed that the reader has a knowledge of semiconductor devices and circuit theory approaching that of a first degree in electronic engineering. In presenting the work, the aims have been to provide a useful text for final year undergraduates specializing in semiconductor device engineering, while the whole treatment should be suitable for a master's degree student or a professional engineer who requires an introduction to charge-transfer devices. Many references are provided in order to lead the reader to greater detail if he is to become involved with the practice of the art.

In a subject that combines several areas of expertise, it is inevitable that some symbols will have multiple use if the text is to be compatible with the more detailed developments of the published papers. Care has been taken to avoid overlap where confusion could arise and the context indicates which use of a particular symbol is relevant when the multiple use has been allowed to remain.

Acknowledgements

The author would like to thank J. Carver, A. Chowaniec and R. C. Tozer for the many useful discussions during the course of our studies of charge-transfer devices. He would like to express his gratitude to Mrs B. Cowell and Mrs D. Grayhurst for their painstaking contribution in converting his jumbled scribbling into a readable manuscript through the medium of their typewriters. Finally, he must express his thanks for the patience shown by his family, Pauline, Philip and Julia, during the painful hours of writing.

G.S.H.

Contents

List of Symbols

List of Symbols

a = binary digit
A = electrode area
= loop gain
= primary well charge
= signal amplitude
A_n = charge loss
A_Q = charge gain
$A(\omega)$ = attenuation
b = binary digit
= electrode width
B = correlation pattern amplitude
= loop gain
= secondary smeared charge
B, B_T, BW = bandwidth
C = capacity
= scaling constant
= tertiary smeared charge
$C_1, C_2, C_3, C_a, C_{cc}, C_d, C_g, C_i$ = capacities
C_{dep} = depletion layer capacity per unit area
C_{dg} = drain/gate capacity
C_{ds} = drain/source capacity
C_{out} = output capacity in BB
C_{ox} = insulator capacity per unit area
C_s = input diode capacity
$C_{\Delta, dep}$ = depletion layer incremental capacity per unit area
d = insulator width
D = diffusion constant
D_i, D_s = reset diodes in BB
D_I = electric displacement in insulator
D_n = electron diffusion constant
D_p = hole diffusion constant
D_S = electric displacement in semiconductor
e = electronic charge
E = electric field
E_2 = interface electric field
E_b = fringing, or built-in, electric field
E_c = bias energy per well

E_d = repulsive electric field during charge transfer
E_I = insulator electric field
E_{max} = maximum electric field in buried channel
E_n, E_p = interface electric field in buried channel
E_s = interface electric field
$E_{s, max}$ = maximum interface electric field
E_{-u} = electric field within stored charge
f = frequency
f_c = sampling frequency
$f_{Nyquist}$ = Nyquist frequency
$f(t)$ = temporal response
$f(\omega)$ = frequency response
FL = field smearing fraction
F_k = Fourier coefficient
F_K = Fourier component in sampled-data format
g = electrode gap
g_m = FET transconductance
g_p = processing gain
G = loop gain
h_k = tap weight
h_m = split electrode dimension
= tap weight
$H(s), H(Z), H(\omega)$ = transfer functions
i = drain or collector current
i_0 = reverse current of p–n junction
J_{gen} = leakage current density
k = Boltzmann's constant
K = integer
= loop gain
K_i = tap weight
l = depletion width
$l_1, l_2, l_b, l_{b,o}, l_{b,o,o}, l_{b,s}, l_c, l_{c,o}, l_{c,o,o}, l_{CD}$ = buried channel depletion widths
L = charge loss fraction
L_i = tap weight
m = integer
m^* = electron effective mass
M = number of delay columns in two-dimensional array
= number of Fourier coefficients in transversal filter
= number of repeated units in BB
n = electron density
= integer
n_i = intrinsic electron density
n_o = electron density boundary value
N = number of delay stages
= number of particles
$\langle N(0)^2 \rangle, \langle N(f)^2 \rangle$ = mean square number fluctuation

N_A = acceptor density
N_C = conduction band density of states
N_D = donor density
N_S = number of charges per store
N_{SS} = surface state density
p = charge fraction
= hole density
= integer
= number of clock phases
p_0 = undepleted hole density
P = arbitrary voltage amplitude
= binary digit
= clock power dissipation
P_{SPS} = power dissipation in series–parallel–series store
q, q_s = signal charge
q_B = charge on electrode
q_{in} = fill and spill input charge
Q = BB signal charge
= Q-factor
Q_{ds} = charge deficit on drain-source capacity
Q_s, Q_{s_2} = charge deficit
$Q_{s,f}$ = terminating charge deficit
$Q_{s,max}$ = maximum charge deficit
$Q_{s,0}$ = initial charge deficit
$Q_{s,r}$ = residual charge deficit
r = integer
= spatial coordinate
R, R_s = resistances
Re = real part
s = complex frequency variable
S_A, S_B, S_C = output signals
S/N = signal to noise ratio
t = time
= depletion layer length increment
T = absolute temperature
T_1, T_2 = transistor switches
T_c = clock period
T_d = total delay in CTD
T_e = emitter-follower transistor of BB
T_i = input sampling transistor of BB
T_s = characteristic time for self-induced charge transfer
T_T = charge-transfer time
T_γ = characteristic time for charge deficit transfer
u = spatial coordinate
v = mean thermal speed
v_{th} = r.m.s. thermal velocity

V = electrostatic potential
$V_{1,2}$ = base-emitter or gate-source voltage
$V_{1,3}$ = collector-emitter or drain-source voltage
V_A = a voltage constant
V_b = base voltage
V_{b_1}, V_{b_2}, V_{c_1}, V_{c_2} = voltages at bases and collectors of a bipolar BB
V_B = bias voltage
V_c = clock amplitude
= collector voltage
V_{ds} = drain-source voltage
V_{DC} = d.c. bias voltage
V_{FB} = flat band bias voltage
V_{gs} = gate-source voltage
V_H = potential hump height
V_{in} = input voltage
V_I = electrostatic potential in insulator
V_{ID} = input diode voltage
V_{Imax} = input voltage range
V_{G_1} V_{G_2} = input gate voltages
V_m = = buried channel peak potential
V_n = buried channel surface potential
= nth output voltage
$V_{n,in}$ = input noise voltage
$V_{n,\, out}$ = output noise voltage
V_o = clock amplitude
V_{out}, $V_{out,\, m,\, u}$, $V_{out,\, m,\, 1}$ = output voltages
V_p = buried channel voltage at semiconductor interface
V_R = input diode clamping voltage in capacitive metering technique
V_s = interface or surface potential
V_s, V_{s_1}, V_{s_2}, V_{s_3}, V(t) = signal voltages
$V_{s,0}$ = surface potential with no charge
V_{S1} = equilibration surface potential; BB signal
V_{SH} = surface potential in electrode gap
V_t = maximum buried channel bias voltage
V_T = threshold voltage
V_{-u} = voltage across stored charge
V_w = voltage to remove potential hump
V_x = voltage at output tapping point of BB
w, w_G = electrode length
w = spatial coordinate
$w(n\tau)$ = data sequence
x = spatial coordinate perpendicular to interface
$x(n\tau)$ = data sequence
$X(Z)$, $X_N(Z)$, $X_S(Z)$ = Z-transforms
y = spatial coordinate parallel to interface
$y(n\tau)$ = data sequence

Z = delay operator

α = attenuation constant
β = current gain of bipolar transistor
$\beta(\omega)$ = phase constant
γ, γ_0 = MOSFET transconductance constants
δ = charge-transfer fraction
$\Delta_i, \Delta^1{}_i$ = smeared signal components
$\Delta\ell_b, \Delta\ell_c$ = increments of buried channel depletion widths
ΔN_{tot} = total transfer noise in BB
$\langle\Delta N^2\rangle, \langle\Delta N_p^2\rangle$ = mean square number fluctuation
Δq = an increment of stored charge
$\langle\Delta q^2\rangle$ = mean square charge fluctuation
Δq_o = increment of charge available for transfer
Δq_r = increment of incomplete charge transfer
$\Delta q_G, \Delta q_s, \Delta q_{s,a}, \Delta q_{s,b}$ = charge increments on a floating gate
ΔT = output pulse length
$\langle\Delta V^2\rangle$ = mean square voltage fluctuation
$\Delta V_1, \Delta V_2, \Delta V_3, \Delta V_B, \Delta V_{BN}', \Delta V_{FG}$ = increments of electrode voltage
ΔV_m = increment of buried channel peak potential
ΔV_{in} = increment of input voltage in fill and spill mode
ΔV_{out} = increment of output voltage in fill and spill mode
$\Delta\rho_n$ = increment of stored charge per unit area
ϵ = remaining charge fraction
$\epsilon(t)$ = a.c. energy
ϵ_I = insulator relative permittivity (2.8)
ϵ_0 = permittivity of free space ($8.85\ 10^{-12}$ F m^{-1})
ϵ_s = semiconductor relative permittivity (11.3)
λ = matched filter tap coordinate
= mean free path
$\lambda_n, \lambda_{no}, \lambda_{ns}$ = minority charge layer thickness
μ = charge carrier mobility
= linear frequency modulation constant
μ_n = electron mobility
μ_p = hole mobility
ξ = surface state energy level
ξ_c = conduction band energy
ρ = charge per unit area
ρ_{dep} = depleted charge per unit area
ρ_F = fixed charge per unit area
ρ_n = mobile charge per unit area
$\rho_{n,m}$ = mean mobile charge per unit area
$\rho_{n,m,0}$ = initial average mobile charge per unit area
ρ_{n0} = initial mobile charge per unit area
$\rho_{n,p}$ = peak mobile charge per unit area
$\rho_{n,sat}$ = saturation mobile charge per unit area

ρ_{sat} = saturation charge per unit area in buried channel
ρ_{-u} = a part of the mobile charge per unit area
σ = capture cross-section
= real part of complex frequency variable
τ = delay per stage
= transfer period
τ_b = charge-transfer time constant for fringing fields
τ_c = charge-transfer time
= temporal extent of signal
τ_{CD} = mean time between collisions
τ_{Diff} = charge transfer time constant for diffusion
τ_{empty} = surface state release time constant
τ_{fill} = surface state filling time constant
τ_r = recombination time constant
τ_R = charge-transfer time constant for mutual repulsion
τ_s = charge-transfer time constant for self-induced drift
τ_T = coherence time
$\tau_\gamma, \tau_{\gamma f}, \tau_{\gamma o}, \tau_{\gamma s}$ = charge deficit transfer time constants
ϕ = clock phase
= phase constant
ω = angular frequency
ω_0 = cut-off frequency
ω_1 = chirp frequency limit
ω_r = resonant frequency

1

Charge-Transfer Devices (CTD)

1.1 Introduction

The human animal has not been content with the limitations of direct interaction with his environment provided by his senses. He has wished to see, hear or otherwise detect events which are freely occurring or responding to his stimulus but which are beyond his immediate perception. This curiosity has led to the birth of signal processing. Emissions from the event of interest must be transformed so that the senses may be activated and a pattern of events may be recognized. In the electronic or electrical engineering world this extension of the senses has been provided by frequency-domain filtering, for example, where signals with an undesirable repetitive rate are separated from desirable signals by low-, high-, or band-pass filters and related devices.

Another separation criterion can be set up in the time domain by examining signals at different times and recognizing or rejecting temporal patterns among them. The electrical device must store or 'remember' signals for a finite time in order to make the comparisons. Early devices used media which could propagate waves modulated by desired signals and so store them for the finite transit time between input and output. Coaxial transmission lines and acoustic delay lines are examples. An alternative approach is to take proportional samples of a signal at discrete times and place them in isolated stores with the intent of recovering them at a later time. Of course, the process must be repeated more frequently than the rate at which the signal changes. The binary storage system of a digital computer provides one realization of this process.

The objective of this book is to describe another class of devices for time-domain processing; the *charge-transfer devices* (CTD). These provide a realization of *analogue* storage of the isolated, sampled data with a delay between reception and return of signals. This is not to say they cannot be used in digital operation; applications in this capacity will also be described. The sampled signal will be carried proportionately by a quantity of charge.

A simple outline of their *modus operandi* is illustrated in Fig. 1.1. The left-hand switch directs a signal into a store for a fixed time interval. During this time interval the right-hand switch is open. The left-hand switch then opens so that both switches are opened during the following delay interval. The right-hand switch then closes for a fixed time to transmit the signal to the next stage. Connection of a sequence of these elements into a ladder network allows delay of a succession of signal samples and preserves their correct order between the input and output. To avoid mixing of adjacent signal samples during the switching operations it is essential that they are each separated by an empty store.

What is the advantage of discrete sampling over continuous delay systems, such as the electromagnetic and acoustic ones mentioned earlier? The answer lies in the desired delay and its flexibility. Electromagnetic waves travel very fast and either for reasons of bulk or signal attenuation become impractical for delay in excess of about 1 microsecond (300 m at 3×10^8 m s^{-1}). Acoustic waves in stable engineering materials travel at velocities in the range 10^3 m s^{-1} to 10^4 m s^{-1} and

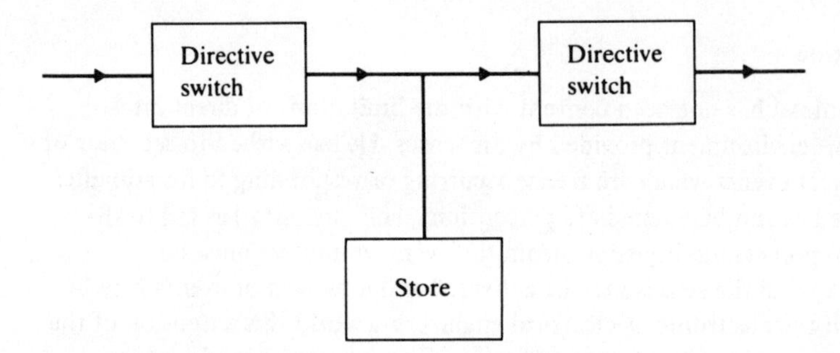

Fig. 1.1 Elements of the analogue, sampled data, delay line.

can provide delays of 100 microseconds, or greater than 1 millisecond with care. Both types of wave have velocities which are not easy to modify, so the delay range available is somewhat inflexible. The advantages of sampled-data delay lines are the longer delay offered which is required by some signal-processing schemes, the flexibility and matching between different devices that is provided by electronic control of the delay time and finally their ability to interact with sensors or complicated signal paths with ease, as will be seen later.

Sensors to extend the visual range and identify spatial patterns are also important. The signal-processing device which detects such patterns may be relatively simple in its function of preserving the correct sequence of the information during conversion to an electrical signal, for example, in the application of *charge-coupled* devices as sensors in television cameras. Moreover, there is the possibility of processing optical signals efficiently at their sensing point.

The objective in the remainder of this chapter is to outline the principles and purposes of charge-transfer devices. The following chapters will provide a more detailed discussion of the various properties, from device design and production to system applications.

1.2 Charge-coupled devices (CCD) and bucket brigades (BB)

The BB directive switch is a bipolar or field effect transistor (FET) and its charge store is a capacitor, C, as illustrated in Fig. 1.2 for an FET realization. One charge sample can be accommodated in a unit comprising two transistors and two capacitors so that neighbouring charges are never connected during the charge-transfer sequence. Alternate transistors have common connections to their gate electrodes (or to the base in a bipolar realization). In essence, the first part of the charge-transfer process is carried out by driving the transistors connected to ϕ_1 in Fig. 1.2 into conduction while those connected to ϕ_2 are non-conducting. In this way charge is transferred from the source capacitor to the drain capacitor of the conducting transistor. The second half of the transfer process is carried out in the same way except that transistors connected to ϕ_2 are conducting and those connected to ϕ_1 are non-conducting. After this second

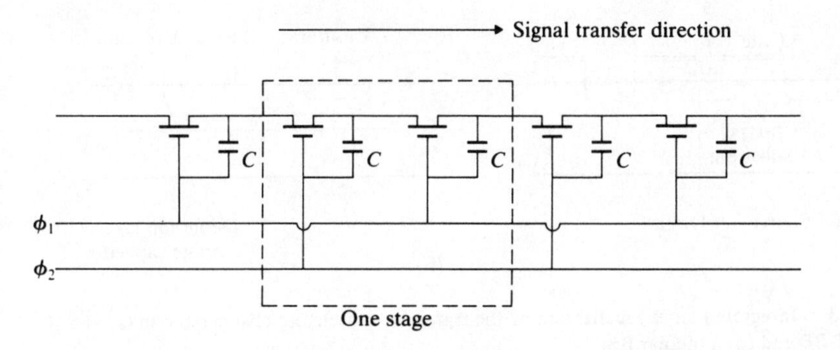

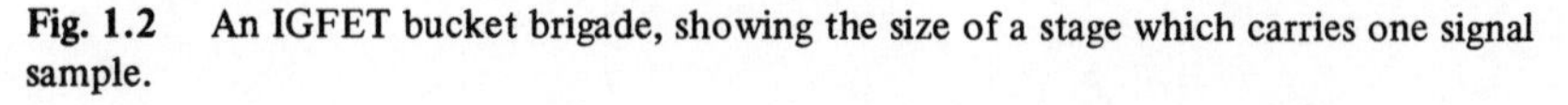

Fig. 1.2 An IGFET bucket brigade, showing the size of a stage which carries one signal sample.

event each analogue signal sample has been advanced one complete stage through the device and the process is cyclically repeated. The delay time is simply controlled by the switching rate of ϕ_1 and ϕ_2. The total delay through the device is the product of the number of samples contained in the device with the time required for each complete switching sequence.

The circuit configuration shown in Fig. 1.2 is only one of many ways of achieving an analogue sampled-data delay but to be useful a circuit must be technologically compatible with integrated-circuit processing. This is necessary in order to build economically the several hundred stages that may be required in any one device. The circuit layout of Fig. 1.2 is particularly advantageous from this point of view. The drain to gate capacity (or collector to base capacity of a bipolar transistor) which is usually an embarrassment in circuit design, is enlarged and turned to good use as the storage capacitor. As shown in Fig. 1.3 this is achieved in an IGFET by allowing the gate and its insulating oxide to overlap the drain while in the bipolar transistor the lateral extent of the base and collector junction is increased so that the storage capacity is provided by the reverse-biased base-collector depletion region. In addition the drain diffusion of one FET is also the source diffusion of the following FET. Closer examination of the charge-transfer process in this type of structure shows that its satisfactory termination requires the signal to be carried as a deficit below a predetermined level. A discussion of these details is deferred until Chapter 3.

Signal entry to the BB is controlled by the gating action of a transistor switch and the output is usually provided by source follower or emitter follower derived connections. Again, full details will be described in Chapter 3. The output can be sensed with little loss of the stored charge at any charge store along the BB so that simultaneous comparisons can be made between signals at successive time intervals. This ability of the BB to act as a tapped delay line allows considerable signal processing power to be brought to bear on a signal through the transversal and recursive techniques that are described in Chapter 6.

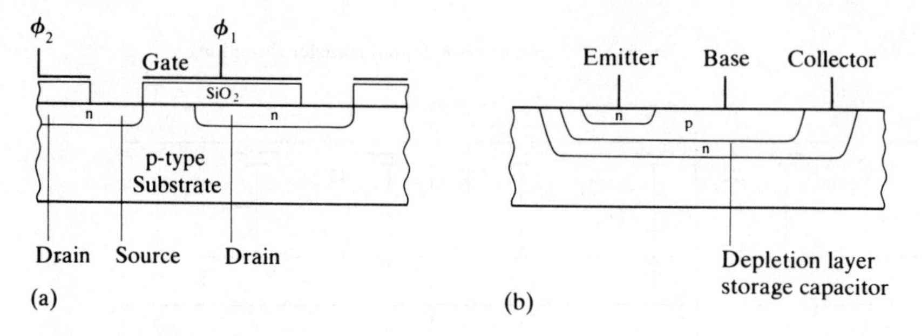

Fig. 1.3 Integrated circuit realization of the transistor switch and charge store in (a) an IGFET BB and (b) a bipolar BB.

Even though the BB has great simplicity and circuit elegance, the 'packaging' of a signal sample and its sequential transfer is carried out far more neatly and efficiently in a CCD. The essential simplicity of the device is illustrated by the cross-section in Fig. 1.4. The construction uses an SiO_2 insulating layer on a p-type silicon substrate. When biased positively, electrodes on the upper surface of the SiO_2 repel the majority holes in the p-type silicon away from the Si–SiO_2 interface leaving a depletion region which is shown by the shaded area of Fig. 1.4. A larger positive bias voltage (on ϕ_2 electrodes in Fig. 1.4) causes a greater penetration of the depletion region into the p-type silicon substrate. The n-type diffusion at the input and the input gate act as the source and gate of an FET which can inject electrons into the 'drain' region under the ϕ_1 electrode. These electrons are minority carriers in the p-type substrate and are attracted towards the positive potential of the electrode above the SiO_2 layer. This attraction is stronger for a higher electrode potential. The minority electrons cannot proceed through the SiO_2 insulator so they collect at the Si–SiO_2 interface where, in effect, there is a potential minimum for them. This potential minimum is deeper for the regions under electrodes with the highest positive bias, so that minority electrons are contained in successive potential minima under one electrode set. Charge transfer in the structure of Fig. 1.4 is achieved in a three-phase fashion.

Increase of the positive voltage on ϕ_3 electrodes until it is equal to that on ϕ_2 electrodes will allow the contained charge to spread under the ϕ_3 electrode region. The positive voltage on the ϕ_2 electrodes is next reduced so that the electron potential minimum only exists under the ϕ_3 electrodes. The process is sequentially repeated from ϕ_3 to ϕ_1, from ϕ_1 to ϕ_2, etc. so that isolated packets of minority carrier charge are transported along the CCD in a sampled-data time delay sequence which is similar to that of a BB. The three-phase sequence is necessary to maintain unidirectional charge transfer and to keep neighbouring charge samples out of contact with each other. Other designs are possible to give two-phase and one-phase operation but these will be described later.

The signal output can be obtained in a destructive manner by 'pouring' the minority charge into the reverse-biased output diode so that it is swept out into

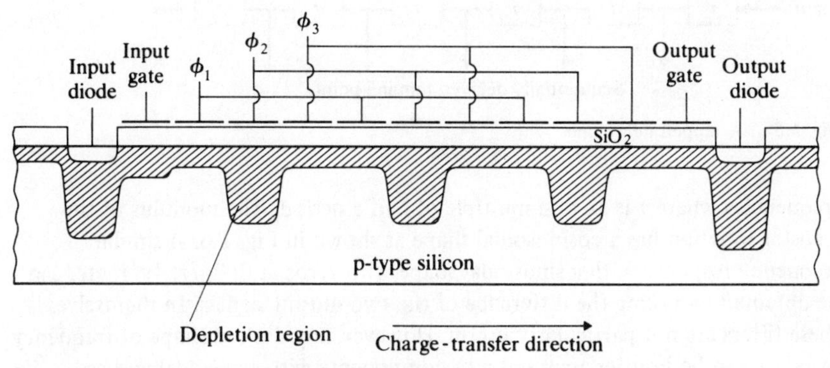

Fig. 1.4 A cross-section of a surface channel CCD.

the external circuit. Multiple tapping points along the CCD which sense charge non-destructively can be implemented by sensing the charge flow to the exposed surface of an electrode. This flow must occur in sympathy with any minority charge flow into the semiconductor region under the electrode. The details of these techniques will be found in Chapters 2 and 5.

One great advantage the CCD has over the BB is its largely fluctuation-free transfer of charge. In the CCD signal charge is essentially 'poured' from one potential well to the next and it has nowhere else to go. In the BB the transfer process is controlled by a transistor which inevitably has undesirable and spurious control signal paths which can allow electrical noise fluctuations. The low noise property of CCD is exploited in low light television cameras as described in Chapter 7 and in computer memories which require as many sequential charge stores as possible as described in Chapter 8.

Even though CCD are made using Metal–Oxide–Semiconductor (MOS) technology they have some requirements which are quantitatively different from those for field effect transistor structures. An outline of these technological advances and their relationship to particular CCD properties is described in Chapter 9.

1.3 Frequency filtering with CTD

CTD provide delay times which are electronically controllable from the sub-microsecond range to approximately one second. In this simple form they can find many uses such as correction of speed fluctuations in audio and video tape recorders and provision of the delay needed in phase alternate line (PAL) colour television receivers. A variety of other applications are described in Chapter 6. However, they are much more versatile as delay lines with a multitude of sequentially delayed parallel outputs as illustrated in Fig. 1.5.

A simple filter can be made in which two output signals are added with delay τ between them. They will destructively interfere with each other at

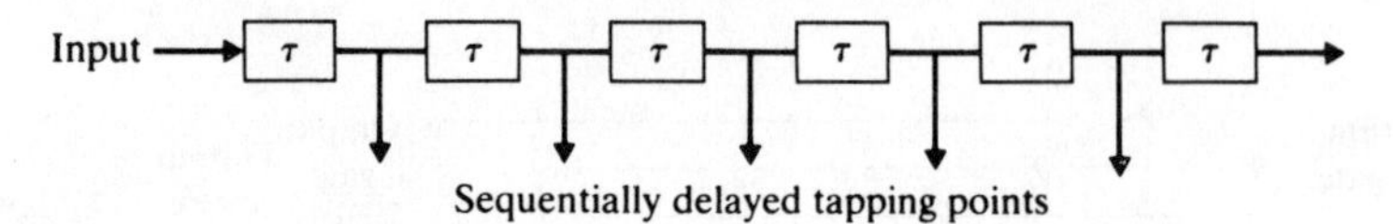

Fig. 1.5 A tapped delay line.

frequencies where τ is an odd multiple of half a period. The modulus of the transfer function has a cosinusoidal shape as shown in Fig. 1.6. A similar frequency response with a sinusoidal shape with zeros at 0, $2\pi/\tau$, $4\pi/\tau$, etc. can be obtained by taking the difference of the two output signals. In themselves these filters are not particularly useful. However, any desired shape of frequency response can be Fourier analysed into components with sinusoidal and co-sinusoidal frequency responses. These can be provided by the tapped delay line of Fig. 1.5, in what is known as a transversal filter. The components with the smallest repeat distance along the frequency axis will require tapping points with the largest number of delays between them.

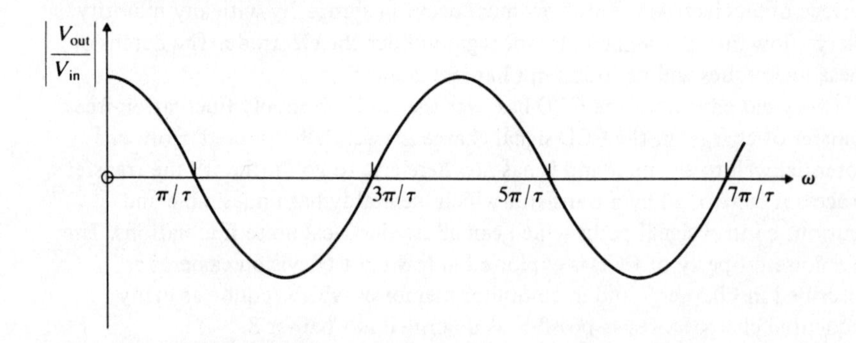

Fig. 1.6 The shape of the transfer function of a simple transversal filter.

If an output from the tapped delay line is fed back to the input a constructive interference will occur when τ is an even multiple of the signal period. Such behaviour causes peaks or resonances in the frequency response of this so-called recursive filter. The recursive filter is the lumped circuit analogue of the optical Fabry–Perot interferometer which has great value in spectroscopy by providing narrow transmission bands.

If the input signal is a short pulse which repetitively occurs at intervals equal to the delay time of a recursive filter, successive pulse inputs will add coherently. Signals of this form occur in radar systems and they are often obscured by noise. In successive coherent additions of the pulses the random noise content only adds in a root mean square fashion so this so-called video integrator will improve the signal/noise ratio of the received signal. Moreover, if the delay time of the delay line contains many CTD segments each one can contain information relevant to radar returns from different ranges so that the cyclic integration process is able to preserve the separate units of information. Greater detail about video integrators will be found in Chapter 6.

When the zeros of transversal filters and the peaks of recursive filters are combined they provide the zeros and poles of frequency responses in a similar way to the behaviour of lumped-circuit filters. This can have great practical value in the design of high quality filters at audio and video frequencies. In a simple description, the phase differences between sinusoidal voltages and currents in inductors and capacitors can be loosely regarded as temporal delays. At low frequencies these lumped components become cumbersome, particularly from the point of view of the integrated-circuit designer. The availability of a 'compact' delay element compatible with integrated-circuit technology and having delay values up to the order of one second is therefore attractive when used with the aid of a modified form of active filter synthesis. The details of the design procedure are described under canonical filters in Chapter 6.

1.4 Matched filtering or pattern recognition with CTD

The tapped delay line in Fig. 1.5 has another powerful signal processing use; to recognize the arbitrary shape or signature of a waveform. The waveform is sampled at suitable intervals and the samples are fed into a CTD with a number of tapping points equal to the number of samples in the pattern length. Each sequential output is effectively multiplied by a constant, or tap weight, at its tapping point by adjustment of the signal gain after sensing. The process is illustrated in Fig. 1.7 where it can be seen that the sum of all the outputs is taken after multiplication by their tap weights. The distribution of tap weights along the CTD has the same shape as the waveform to be recognized. When the waveform completely occupies the CTD its pattern matches that of the tap weights and there is a maximum output or correlation peak from the summing point. This peak essentially occurs at the time of coincidence because negative or positive values in the waveform (relative to the mean level) are everywhere multiplied by negative or positive tap weights respectively. In this way, all the output signals to the summing point have positive values. This is not the case for other relative 'positions' of the waveform and the tap weights, so the combined output signal at any other time will be smaller.

There is another advantage in this detection process. Wherever the signal is small it is multiplied by a small tap weight at the time of coincidence. This implies that electrical noise on the small parts of the waveform is strongly attenuated. Therefore, the process effectively discriminates against noise and improves the signal/noise ratio compared with simpler detection techniques. Correlation detection of this type appears to have many applications both in communications, and in instrumentation, where signals frequently occur with a particular signature. Several detailed examples are given in Chapter 6. One example is of particular note because it allows the Fourier transform of a signal to be obtained at bit rates in excess of 1 MHz.

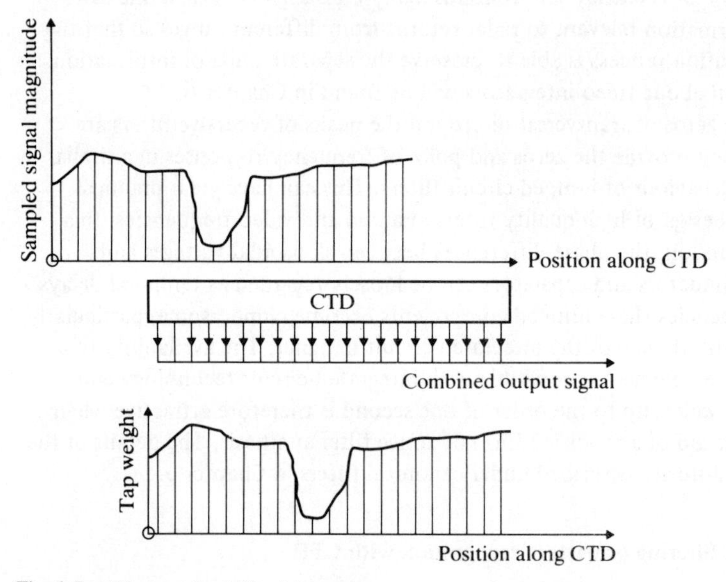

Fig. 1.7 The matched filtering process.

1.5 Imaging with CCD

If a photon is absorbed in the depletion region of a CCD with the generation of an electron–hole pair, the majority carrier moves into the substrate beyond the depletion layer edge and is lost. The minority carrier will be collected in the nearest potential minimum at the $Si-SiO_2$ interface. This optical input process may be used as the basis of a television camera if an optical image is focused on to an array of many parallel CCD with close enough spacing of their potential wells. Minority carriers are generated at each point of the array in direct proportion to the local light intensity and are received and stored in the nearest potential well. The clock sequence which causes charge transfer is started after a quiescent optical reception period. Each line of data can be sequentially transferred out of the device in a manner which is naturally compatible with raster-scan television. The optical integration is then repeated and charge is transferred to the output at the end of each successive frame. The transfer of charge out of the device has to be as rapid as possible, otherwise the continuing optical flux will add a significant number of minority carriers to the transferring charge in the wrong relative position and will be the cause of image blurring.

Alternatively the charge-transfer process could be carried out in potential wells which are shielded from the light but this would cause inefficient use of the silicon chip area. The several possible compromises and different techniques needed to realize a practical television camera are discussed in Chapter 7.

1.6 Computer memory

The relatively small area of the charge storage cell in a CCD is attractive to the computer memory designer who wishes to store quantities of binary data. CCD are not as versatile as random access memories using MOS FET technology because the data must be stored in long sequences which are only accessible on entry and exit from the device. However, they are attractive for situations where large blocks of data must be stored for later sequential transfer to the processing parts of the computer. Their attraction in this situation arises from the simplicity of the CCD and its high storage capability which results in low cost. They are also attractive when compared with the much higher storage capability and lower cost of magnetic disc and drum storage techniques, owing to their much higher operating speed. CCD appear to fill a gap in access, time, cost and storage capacity which has existed between semiconductor random-access memories and the magnetic bulk memory technologies. Several different arrangements of the storage wells and their relationship to practical requirements are discussed in Chapter 8.

1.7 Charge-transfer defects

The first problem introduced in this section concerns input and output circuits. A major source of non-linearity can occur if proper consideration is not given to the fundamental nature of the transduction process. Conventional circuits usually operate with voltages or currents controlling secondary voltages or currents. For example, in a bipolar transistor in linear operation, the base current controls the collector current while the gate-source voltage of a field effect transistor controls its drain current. In charge-transfer electronics the currents or voltages of the external circuitry must be linearly converted to a signal charge which is usually less than 10^7 electrons in a CCD. Charge can be obtained as the time integral of current but such a technique does not usually give good linearity in charge-transfer electronics. New techniques have had to be developed which carry out the conversion process directly; these techniques are described in Chapter 5 for CCDs and in Chapter 3 for BBs.

An inherent problem of both BBs and CCDs is charge-transfer inefficiency. A small quantity of charge is left behind at each transfer event and even though it may be as small as 10^{-5} times the transferred charge it can cause a significant corruption of a following charge sample after several hundred transfer events. In filtering applications it causes a shift of the critical frequencies and attenuation coefficients in addition to non-linear distortion of the signal in extreme cases. In a matched filter, charge-transfer inefficiency will alter the shape of a signal and reduce the efficiency of its recognition. Television pictures suffer image blurring if a significant amount of charge is allowed to enter the wrong storage cell of a CCD camera. In computer memories the maximum length of the store is limited by the accumulating residual charge after many transfers when the recognition of a binary 0 following a binary 1 becomes impossible. Various techniques are available to enable the effects in a particular application to be minimized and they will be described at the appropriate places in the following chapters. However, to produce a widely useful device it is necessary to reduce the charge-transfer inefficiency as much as possible.

One simple cause of incomplete charge transfer is when the transfer rate is too high. A finite time is required to achieve nearly complete charge transfer; the semiconductor transport processes responsible for the transfer are described in Chapter 2 for CCDs and in Chapter 3 for BBs. The ultimate limitation is imposed by defects of the semiconductor in the charge storage and transfer region. In the case of a surface-channel CCD, this is the Si–SiO_2 interface, where the defects can be particularly bad. Spurious and localized energy levels occur in the crystal lattice. They can trap the minority carriers and subsequently release them after one or more transfer events have occurred. A technique popularly known as 'fat zero' can be used to reduce significantly the charge smearing effect of these surface states. In essence the objective is to keep the spurious surface states permanently full of charge by always maintaining the signal level greater than a prescribed zero level. This is typically 20% of the maximum charge that can be stored in a single potential well. Unfortunately this technique is not completely successful and does introduce a form of generation–recombination noise into the CTD. In the case of the CCD, if the lowest possible charge smearing is required it is necessary to move the charge store and transfer channel away from the Si–SiO_2 interface and into the bulk of the semiconductor where the defect density is much smaller. The device is then a buried channel CCD and the physical location of the channel is described in Chapter 2. A full and quantitative discussion of charge-transfer defects is given in Chapter 4.

2

Charge-Coupled Devices

2.1 Introduction

In the CCD version of charge-transfer devices it is possible to identify a signal store but the switching is not provided by a separately identifiable circuit element. Rather, it is carried out by relative manipulation of adjacent charge stores. The store is a Metal–Insulator–Semiconductor (MIS) capacitor usually realized with Metal–Oxide–Semiconductor (MOS) integrated-circuit technology.

2.2 Charge storage in a Metal–Insulator–Semiconductor (MIS) capacitor

A section, perpendicular to the surface, through a MIS capacitor is shown in Fig. 2.1. A positive-bias voltage has been applied to the metal contact and the diagram for electric field, E, implies a p-type semiconductor substrate. The electrons (minority carriers) are attracted towards the insulator owing to the positive bias on the metal, and the holes are repelled towards the substrate contact leaving a region depleted of mobile charge carriers. For the moment diffusion will be neglected and the semiconductor is assumed to be uniformly doped with singly-charged acceptors of density, N_A.

In the semiconductor, Poisson's equation in one dimension gives us

$$\epsilon_s \epsilon_0 \frac{\partial E}{\partial x} = e(p - n - N_A). \qquad 2.1$$

where n is the free-electron density, p is the free-hole density, and e is the modulus of the electronic charge. In the undepleted region $p - n = N_A$ so that $\partial E/\partial x = 0$. In the depletion region equation 2.1 reduces to

$$\epsilon_s \epsilon_0 \frac{\partial E}{\partial x} = -N_A e \qquad 2.2$$

If the edge of the depletion region is chosen as the spatial origin (where $E = 0$ to give zero conduction current) integration of equation 2.2 gives the variation of E with x in the depletion region:

$$\epsilon_s \epsilon_0 E = -N_A e x \qquad 2.3$$

The difference of the electric displacement D on either side of the insulator–semiconductor interface is equal to the stored charge per unit area, ρ,

i.e. $$D_I = D_S + \rho \qquad 2.4$$

The subscripts refer to insulator and semiconductor respectively and ρ may be

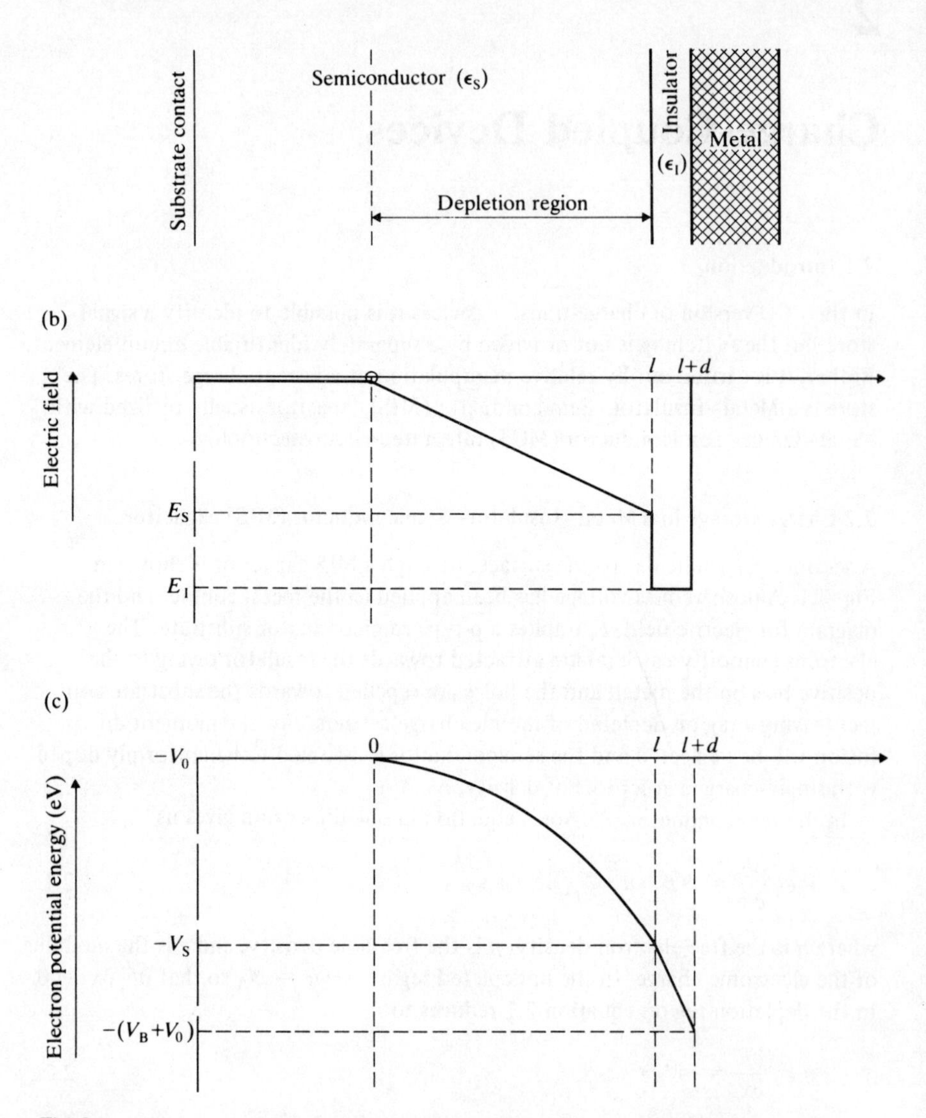

Fig. 2.1 The cross-section of an MIS capacitor and its associated electrostatic conditions.

fixed charge, ρ_F, caused by ionized impurity atoms or defects of the semiconductor crystal at the interface, or it may be due to mobile minority carriers, ρ_n, which we will see can be stored in this region.

From equation 2.3 the depletion layer width, l, is related to the interface electric field, E_S, in the semiconductor by

$$\epsilon_S \epsilon_0 E_S = -N_A e l \qquad 2.5$$

and equation 2.4 may be written as

$$\epsilon_I\epsilon_0 E_I = \epsilon_s\epsilon_0 E_s + \rho$$
$$= -N_A el + \rho \qquad 2.6$$

In the ideal situation where there is an absence of any surface charge, the ratio E_I/E_s is equal to the inverse of the dielectric permittivity ratio. If the insulator is ideal and has no fixed charge in its bulk, E_I will be constant.

The potential at any point is $V - V_0 = -\int E\mathrm{d}x$ where V_0 is the potential of the substrate for $x < 0$ and is different by the bandgap energy for electrons and holes. Many of the following discussions will use the minority carrier potential energy, and in particular its minimum value, in order to describe the control of stored charge. In the case of a p-type semiconductor, the voltages will all be positive with respect to the substrate, so that the potential energy of the minority electrons, expressed in electron-volts, will simply be the negative of the electrostatic potential.

It is the electron potential energy which will appear on diagrams of electrostatic conditions (e.g. Fig. 2.1(c)) so that the hole potential energy will decrease upwards. Care must be taken to avoid confusion with this sign difference for potential and potential energy because the general literature of MOS devices often describes a potential energy simply as a potential.

In the case of an n-type substrate, the voltages would all be negative with respect to the substrate, the minority carriers would be holes, so that their potential energy in electron-volts would have the same sign as the electrostatic potential.

In the depletion region, using equation 2.3,

$$(V - V_0) = -\int E\,\mathrm{d}x$$
$$= \frac{N_A e}{\epsilon_s\epsilon_0}\frac{x^2}{2} \qquad 2.7$$

$$\therefore \quad V_s = \frac{N_A e}{\epsilon_s\epsilon_0}\frac{l^2}{2} + V_0 \qquad 2.8$$

The interface potential energy, $-(V_s - V_0)$ is usually referred to as the surface potential. In the insulator of width, d, using equation 2.6

$$V_I = V_s - E_I(x - l) \qquad (l < x < l + d) \qquad 2.9$$

$$\text{i.e.} \quad V_I = V_s + \left(\frac{N_A el - \rho}{\epsilon_I\epsilon_0}\right)(x - l) \qquad 2.10$$

$$\text{or} \quad V_I = V_0 + \frac{N_A e}{\epsilon_s\epsilon_0}\frac{l^2}{2} + \left(\frac{N_A el - \rho}{\epsilon_I\epsilon_0}\right)(x - l) \qquad 2.11$$

The bias voltage appearing across the insulator and semiconductor is V_B given by

$$V_B = \frac{N_A e}{\epsilon_s\epsilon_0}\frac{l^2}{2} + \left(\frac{N_A el - \rho}{\epsilon_I\epsilon_0}\right)d \qquad 2.12$$

It was implicitly assumed above that the electron could exist at any point in space but this is not true. Its existence depends on the value of V_0 relative to the band structure energies of the semiconductor or insulator. In Fig. 2.2 two cases are shown where V_0 corresponds to the energy of an electron or of a hole at the respective band edges of the isolated semiconductor. Also shown are the allowed energy levels of the insulator and the metal. Any electrons falling down the

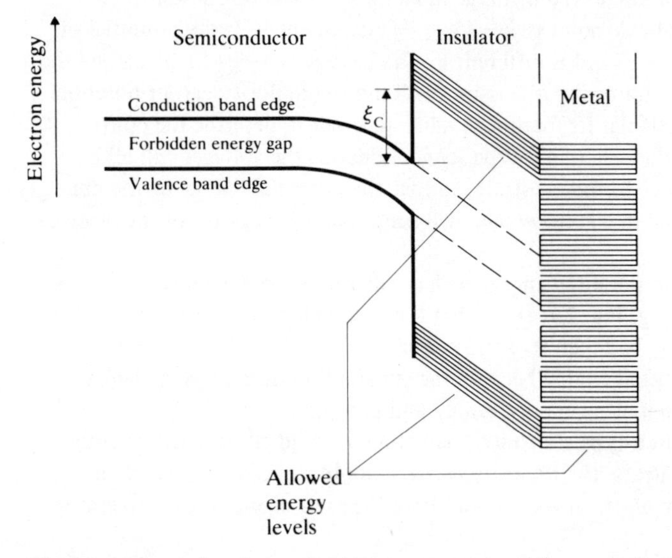

Fig. 2.2 The semiconductor forbidden energy gap and the energy levels in the insulator and metal of an MIS diode.

potential slope in the conduction band cannot proceed past the insulator interface. If the insulator is a good one the value of ξ_C (in Fig. 2.2) needs to be much greater than a few times kT so that an insignificant number of electrons are excited over the potential barrier. Accordingly, there is a potential minimum for the minority electrons at this interface allowing them to be stored out of contact with the majority holes, so avoiding recombination. A free hole would rise up the electron energy scale along the valence band-edge contour and away from the semiconductor–insulator interface.

If the potential at the interface is sufficiently large, the Fermi level becomes closer to the conduction band than the valence band, or as is usual in a CCD, the Fermi level lies within the conduction band at the interface. In this case the interface is in an *inverted* condition, so that it would contain more electrons than holes if thermal equilibrium were established. The nearly complete removal of holes in the inverted condition is necessary to largely remove the possibility of their recombination with stored minority electrons.

For CCD operation the thermal equilibrium must not be established and the only minority carriers stored at the interface must be those deliberately introduced by processes described later. Accordingly, the maximum storage time is limited by the thermal generation of minority carriers in the depletion region, or by other spurious leakage effects. In practice this time limit approaches one second at 300 K.

All the above arguments would have been equally valid for an n-type semiconductor with $-N_A$ replaced by N_D, the donor density, so that the signs of the field and potential were reversed and holes could be stored at the semiconductor–insulator interface.

Numerical values of several useful parameters relevant to a p-type silicon device with a silicon dioxide insulator are illustrated in Figs. 2.3 to 2.6, as a function of the surface potential $-(V_S - V_0)$. In all these numerical results $\epsilon_S/\epsilon_I = 4$, $\epsilon_S\epsilon_0 = 10^{-10}$ F m^{-1} and $d = 10^{-7}$ m. In Fig. 2.3 the depletion layer width is given from equation 2.8 by

$$l = [2\epsilon_S\epsilon_0(V_S - V_0)/N_A e]^{\frac{1}{2}} \qquad 2.13$$

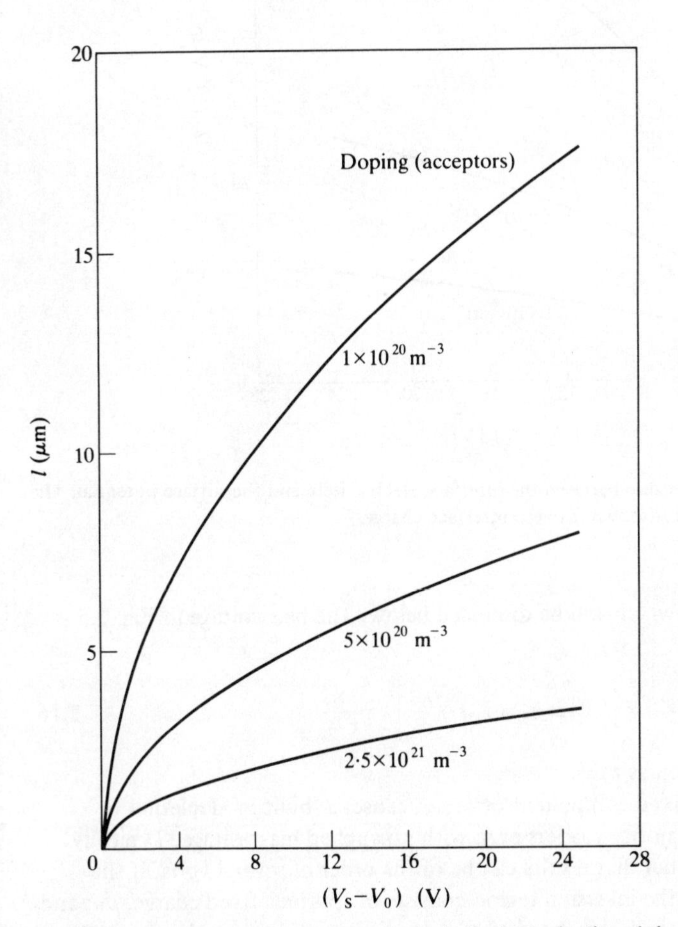

Fig. 2.3 The relationship between the depletion layer depth and the surface potential.

The electric fields at the interface in the semiconductor and in the insulator are given respectively by

$$E_S = -[2(V_S - V_0)N_A e/\epsilon_S\epsilon_0]^{\frac{1}{2}} \qquad 2.14$$

and $$E_I = E_S(\epsilon_S/\epsilon_I) + (\rho/\epsilon_I\epsilon_0) \qquad 2.15$$

In Fig. 2.4 ρ is taken as zero but a zero correction can be added to E_I using the

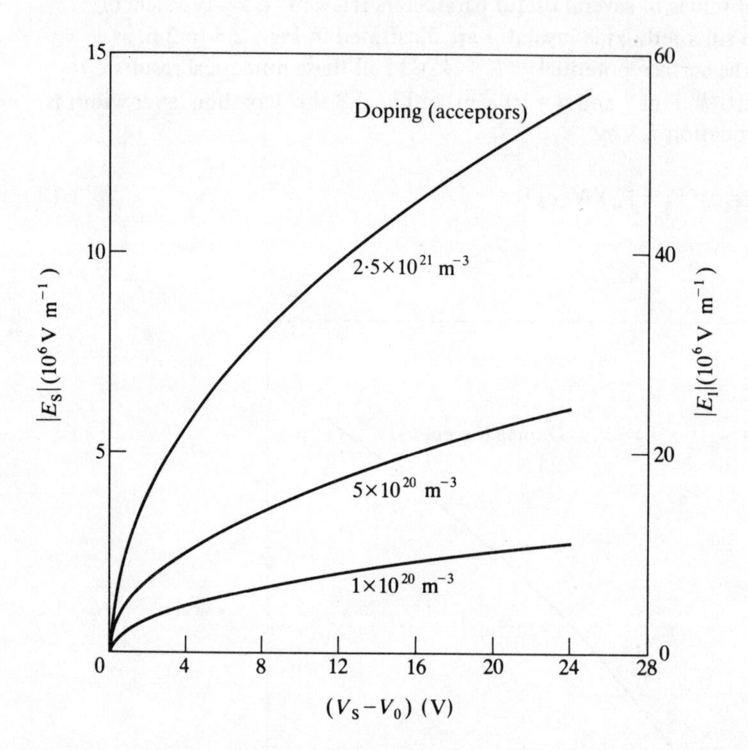

Fig. 2.4 **The relationship between the interface electric field and the surface potential. The insulator electric field is shown for zero interface charge.**

data from Fig. 2.6 which will be discussed below. The bias voltage in Fig. 2.5 is given by

$$V_B = (V_S - V_0) + \frac{\epsilon_S}{\epsilon_I} d\left[2(V_S - V_0)\frac{N_A e}{\epsilon_S\epsilon_0}\right]^{\frac{1}{2}} - \frac{\rho d}{\epsilon_I\epsilon_0} \qquad 2.16$$

and ρ has been taken as zero.

The undesired fixed component of ρ, ρ_F, causes a 'built-in' depletion or accumulation of majority carriers even with no applied bias voltage. Typically the 'built-in' potential that results can be of the order of several volts. If the interface charge in the inversion region consists of spurious fixed charge, ρ_F, and mobile charge, ρ_n, the bias voltage required to remove this potential (i.e. make

$V_s - V_o = 0$ with $\rho_n = 0$ in equation 2.16) is $-\rho_F d/\epsilon_I\epsilon_o$. It is often referred to as the 'flat-band' voltage, V_{FB}. Equation 2.16 is often written in the form

$$(V_B - V_{FB}) = (V_s - V_o) + \frac{\epsilon_s}{\epsilon_I} d \left[2(V_s - V_o)\frac{N_A e}{\epsilon_s \epsilon_o}\right]^{\frac{1}{2}} - \frac{\rho_n d}{\epsilon_I \epsilon_o} \qquad 2.17$$

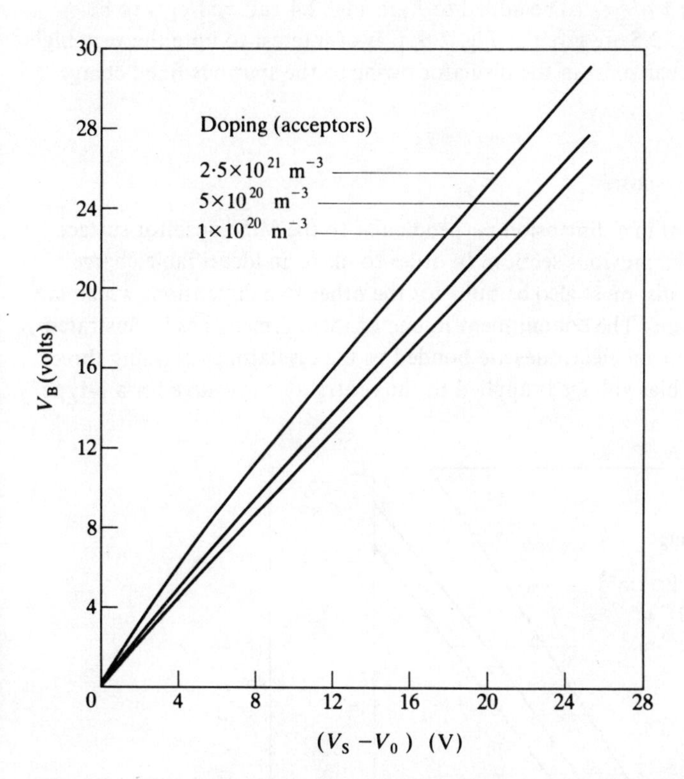

Fig. 2.5 The relationship between the bias voltage and the surface potential.

Alternatively the relationship may be rearranged to give the surface potential $(V_s - V_o)$ as a function of the bias voltage:

$$(V_s - V_o) = (V_B - V_{FB}) + \frac{\rho_n d}{\epsilon_I \epsilon_o} + V_A - \left\{2V_A\left[(V_B - V_{FB}) + \frac{\rho_n d}{\epsilon_I \epsilon_o}\right] + V_A^2\right\}^{\frac{1}{2}} \qquad 2.18$$

In equation 2.18,

$$V_A = \frac{N_A e\, \epsilon_s \epsilon_o}{(\epsilon_I \epsilon_o / d)^2} \simeq 0.3 \text{ V}$$

for $N_A = 10^{21}\,\text{m}^{-3}$ so that the square root term is usually negligible for the working voltages of a practical CCD. In devices made on p-type silicon this fixed charge usually causes depletion (and an n-type inversion channel at the semiconductor–insulator interface) so that it aids any applied bias. To place these results in perspective it may be noted that 10^{-3} C m^{-2} is equivalent to 1 pC of fixed charge under a typical electrode of dimensions 10 μm x 100 μm. This charge is comparable with the saturation value of mobile charge later calculated from equation 2.26 for the data of Fig. 2.3.

The zero errors $\pm\,\rho/\epsilon_I\epsilon_0$ to be added to E_I in Fig. 2.4 and $\pm\rho d/\epsilon_I\epsilon_0$ to be added to V_B in Fig. 2.5 are given in Fig. 2.6. It is of interest to note the very high electric fields that can exist in the insulator owing to the spurious fixed charge at the interface.

2.3 The CCD charge store

Charge containment in a dimension perpendicular to the MIS capacitor surface was described in the previous section. In order to make an identifiable charge store, potential 'walls' must also be built for the other two dimensions which are in the interface plane. The containment in one of these dimensions is illustrated in Fig. 2.7. Three metal electrodes are bonded to the insulator so forming three MIS capacitors. A bias voltage is applied to the central one (positive for a p-type

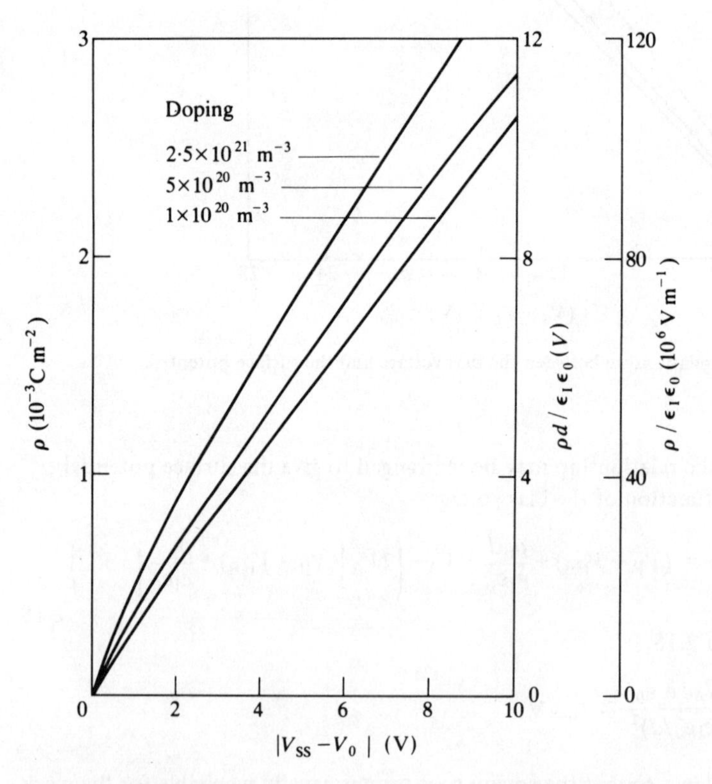

Fig. 2.6 The relationship between the fixed interface charge and the surface potential, $(V_{SS} - V_0)$, for zero bias voltage. The electric field and voltage corrections caused by fixed interface charge are also shown.

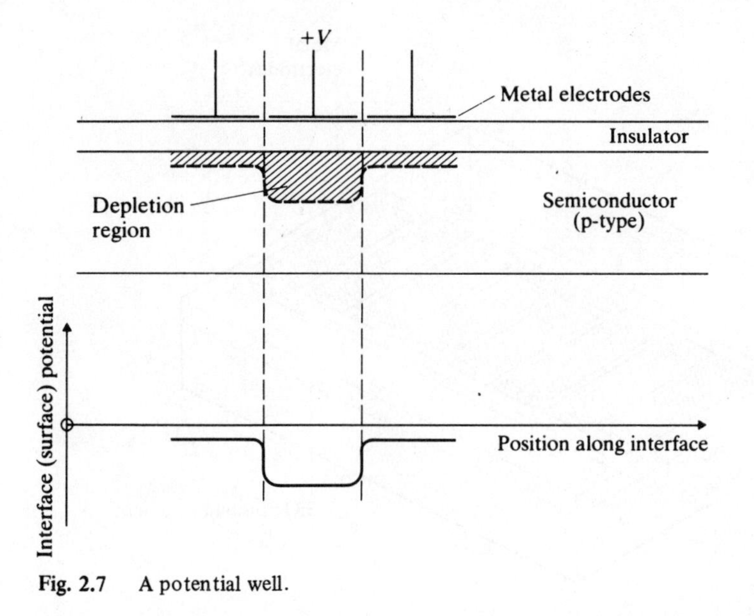

Fig. 2.7 A potential well.

semiconductor) so that a depletion region is formed underneath it in the semiconductor. Correspondingly, the potential at the interface (commonly referred to as the surface potential) is lowered. Any minority carriers present will be contained at the interface and within this potential minimum, providing that the height of the potential walls is significantly greater than the thermal energy, kT.

Containment in the third dimension requires 'potential walls' parallel to the CCD channel. This can be achieved with a p^+-diffusion in a p-type substrate, as illustrated in Fig. 2.8, so that a smaller voltage is dropped across the depletion layer. The p^+-acceptor concentration in this channel stops diffusion and must not be too high, otherwise avalanche breakdown may occur at the interface as described in section 2.8. An alternative way of increasing the proportion of bias voltage dropped across the insulator is to use a thick oxide as illustrated in Fig. 2.9.

2.4 The switching (charge-transfer) process

The CCD is constructed with a large sequence of metal electrodes as illustrated in Fig. 2.8. Fig. 2.10 illustrates one procedure for moving stored charge along this sequence of stores by using a three-phase clock whose voltage variation with time is shown in Fig. 2.11. At time t_1, Fig. 2.10 illustrates the containment of charge under electrodes connected to ϕ_1. At time t_2, bias voltage V_c is applied to ϕ_2, and the contained charge moves by drift and diffusion into the broader containment region under both ϕ_1 and ϕ_2 electrodes. At time t_3, V_c is removed from ϕ_2, relatively slowly, so that the charge-transfer process can be completed. Too rapid a removal of V_c on ϕ_1 before all charge has been transferred may result in some charge flowing in the undesired direction along the 'top' of the surface potential

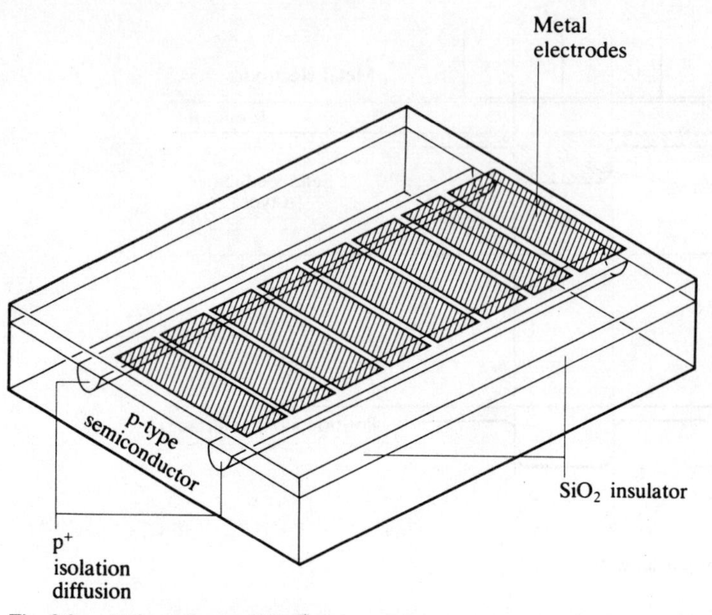

Fig. 2.8 A channel-stop diffusion.

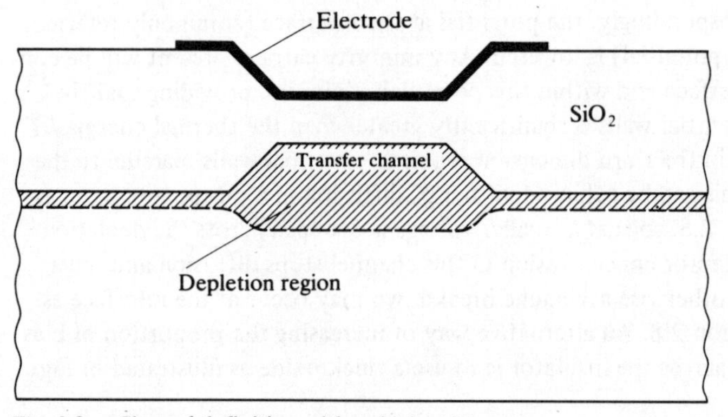

Fig. 2.9 Channel definition with a thick oxide.

profile. ϕ_2 does not have a speed restriction except it should be applied quickly to give as much time as possible for the slow falling edges during the charge-transfer process.

Repetition of this process in successive phases causes sequential transfer of stored minority carriers along the CCD.

2.5 Other clock systems

2.5.1 Two-phase operation

In the previous section it was necessary to use a three-phase clocking system to ensure unidirectional charge transfer. There are other ways of achieving the same

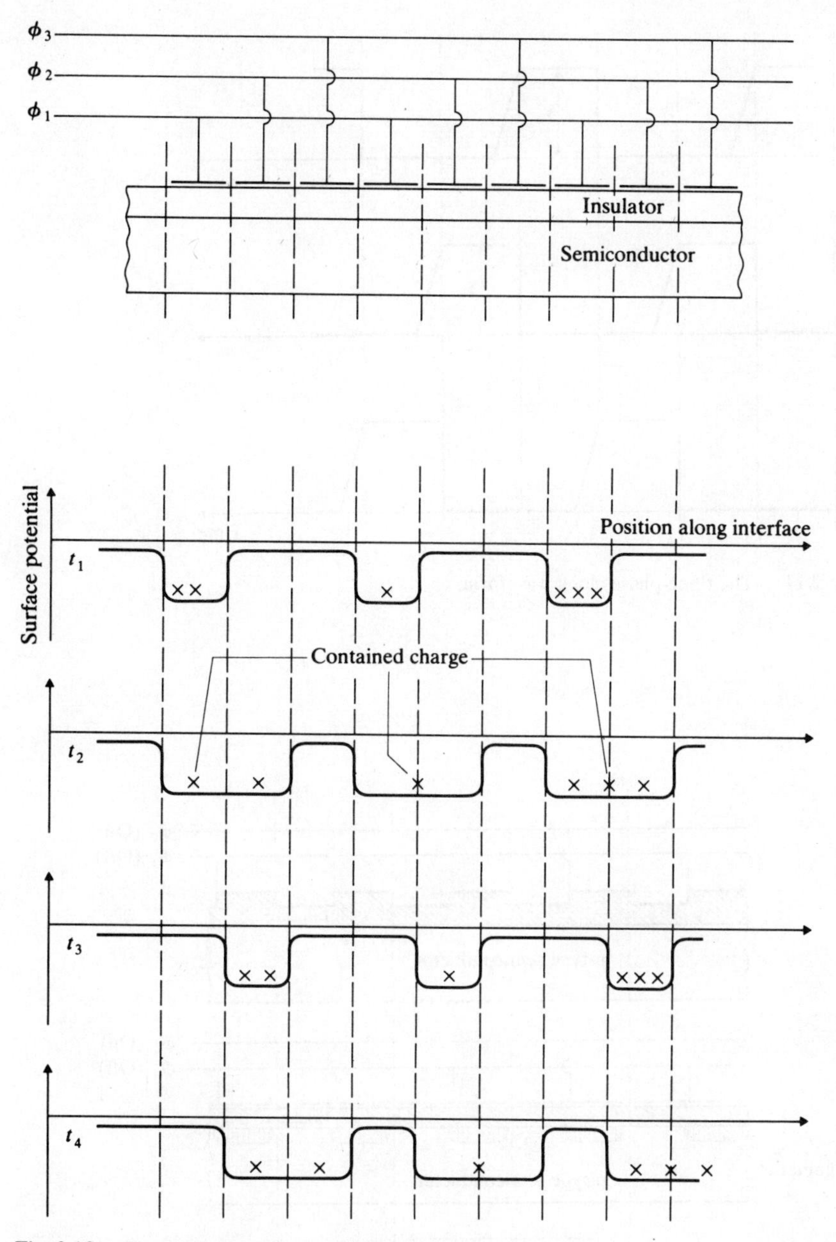

Fig. 2.10 The three-phase, charge-transfer process.

objective. For example, two-phase clocking can be used if the surface potential wells can be constructed with a built-in electric field to ensure the directional property. Two techniques are illustrated in Fig. 2.12. In both cases, the objective is to increase the potential difference across the oxide, over the half of the elec-

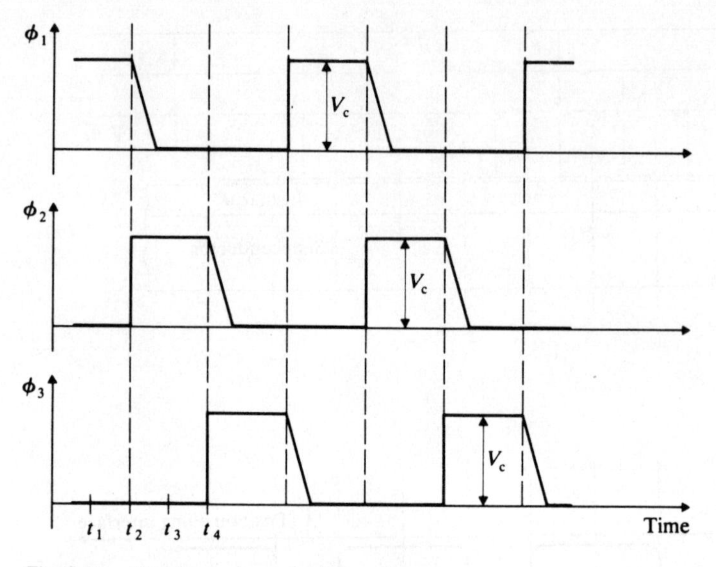

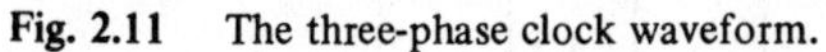

Fig. 2.11 The three-phase clock waveform.

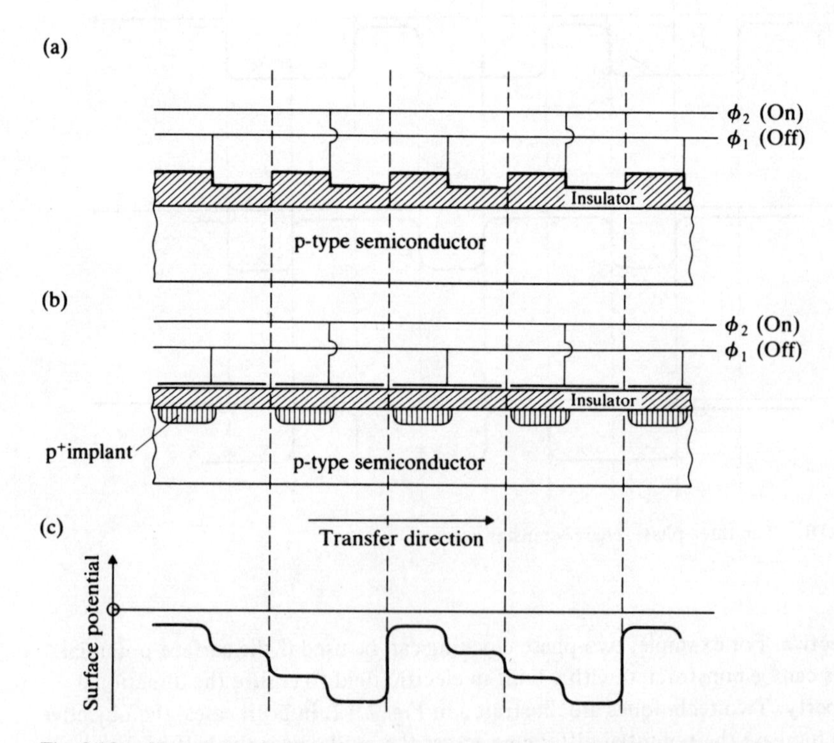

Fig. 2.12 Two-phase clock technology: (a) stepped oxide; (b) diffused implant.

trode closest to the input end of the CCD, so that the corresponding potential difference across the semiconductor is reduced and its potential minimum is not so deep. The stepped oxide technique in Fig. 2.12(a) achieves the objective by increasing the oxide thickness. The diffused implant illustrated in Fig. 2.12(b) uses a more highly doped semiconductor of the same polarity as the substrate to reduce the depletion layer thickness. More design details have been given by Krambeck *et al.* (1971). Reference to Fig. 2.1 and the related discussion will also allow quantitative design guidance. Even though the two-phase system is attractive, both for the relative simplicity it allows in clock-circuit design and for the simplified layout of interconnections on the semiconductor chip, it does have disadvantages. It requires four-thirds of the semiconductor chip area when compared with three-phase clocking because each electrode has two levels of definition. Furthermore there is a reduction of the potential difference (or potential 'wall') between alternate potential wells during the switching process (Fig. 2.13),

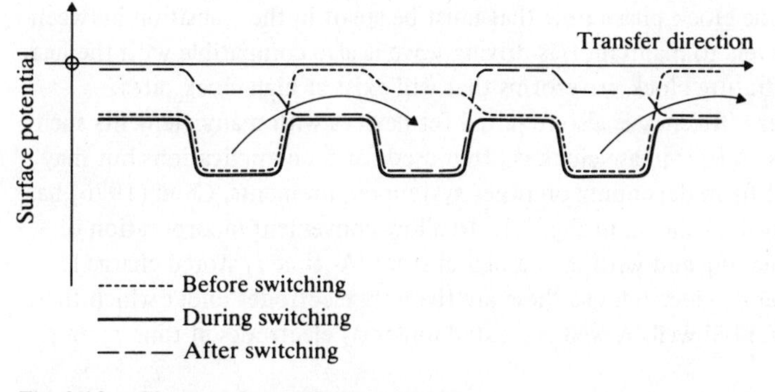

Fig. 2.13 The two-phase switching process.

causing a reduction of the maximum amount of charge that it is possible to contain during transfer when compared with the three-phase device which always maintains a full-height potential barrier on either side of the contained charge in the transfer dimension.

2.5.2 Four-phase operation

The saturation defect of two-phase clocking does not exist in four-phase clocking, there is no increase of chip area per charge containment cell over the two-phase system and the direction of charge transfer can be reversed if required. In essence, the four-phase system provides a double potential barrier between successive charge containment cells, which will help to make the efficiency of charge transfer closer to 100% than in the three-phase system. This may be seen in Fig. 2.14 where the surface potentials are shown at one time when the four electrodes are sequentially clocked so that a potential 'wave' travels along the device. This close concern with high transfer efficiency is of importance (as will be discussed later) when extremely high clocking rates (~100 MHz) are required. The high

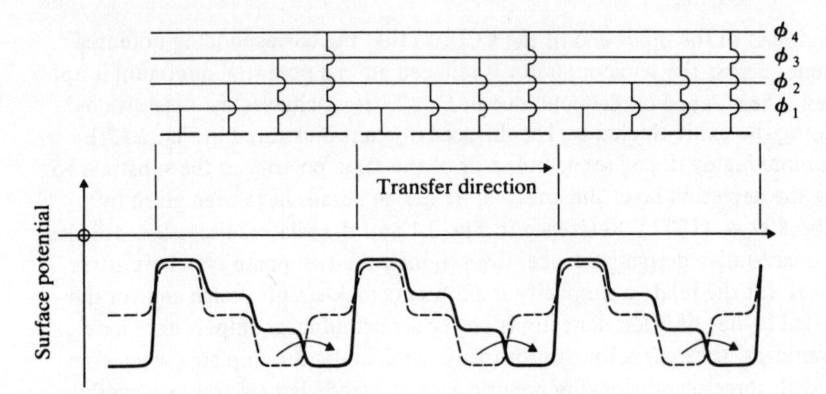

Fig. 2.14 A simple four-phase clock sequence.

proportion of the clock phase time that must be spent in the transition between high and low levels to maintain this driving wave is also compatible with the near sinusoidal quadrature clock waveforms that will exist at high clock rates.

High transfer efficiency is also required for devices with many elements such as digital stores. A four-phase clock is often used for such applications but may take a different form depending on other system requirements. Chou (1976) has described the system shown in Fig. 2.15 to allow convenient incorporation of regeneration, reading and writing in a digital store. At time t_1 stored charge is contained under ϕ_4 electrodes as these are the only electrodes under which there is a surface potential well. A well is created under ϕ_2 electrodes at time t_2, but

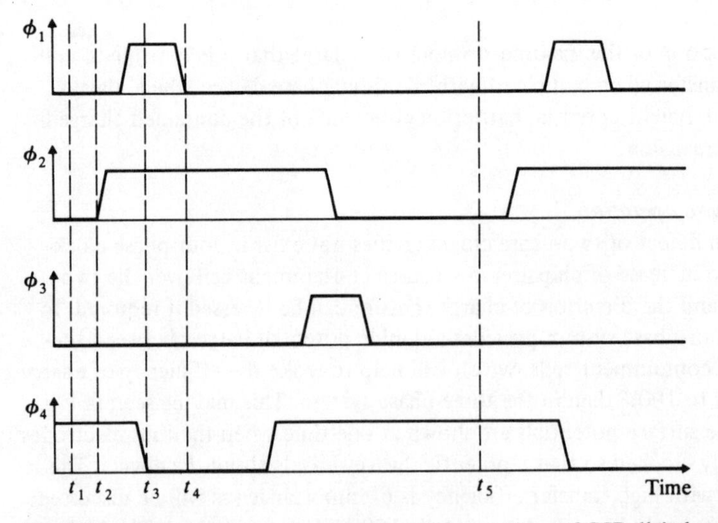

Fig. 2.15 A four-phase clock sequence used for one type of CCD digital memory.

charge cannot transfer owing to the surface potential barriers (in the transfer direction) under ϕ_1 and ϕ_2 electrodes. At time t_3, surface potential wells are created under ϕ_1 electrodes and are removed under ϕ_4 electrodes, so that stored charge resides under ϕ_1 and ϕ_2 electrodes. The potential well under ϕ_1 electrodes is removed at time t_4, so that the stored charge resides under ϕ_2 electrodes. The charge has now transferred over half a complete storage cell of the CCD. A similar sequence is executed for transfer from storage under ϕ_2 to ϕ_4 electrodes and the transfer unit is complete at time t_5.

2.5.3 Pseudo-one-phase and two-phase operation

The driving waveforms for two-phase and three-phase systems can be simplified to one-phase and two-phase respectively if one set of electrodes in each case has a d.c. potential of half the clock driving amplitude (Fig. 2.16). The disadvantage

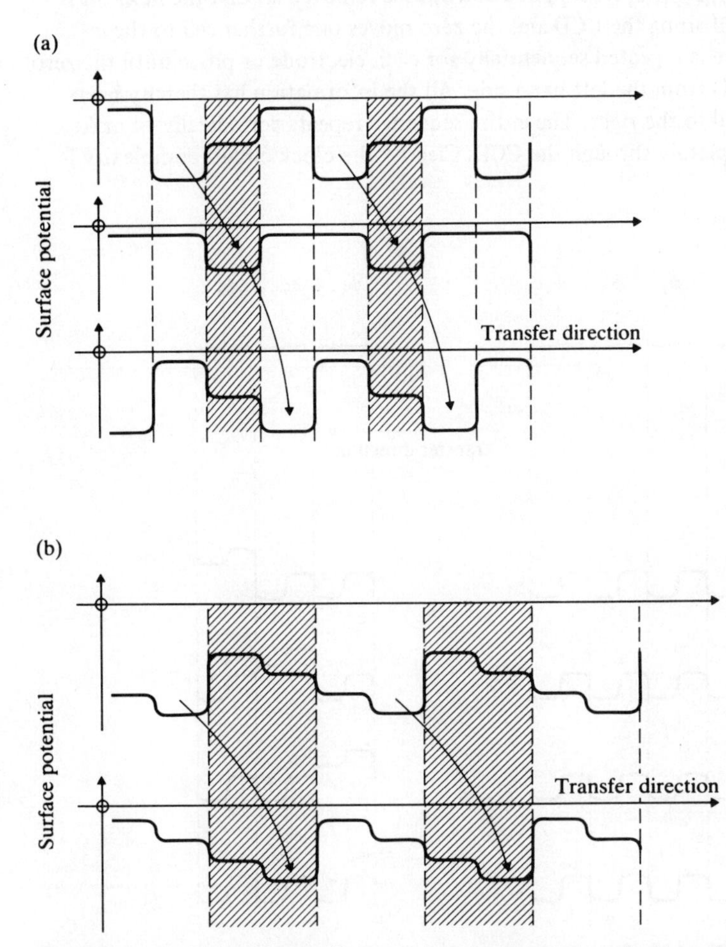

Fig. 2.16 Successive clock cycles for (a) two-phase operation of a three-phase device, (b) one-phase operation of a two-phase device. The shaded regions have constant electrode potential.

of this procedure is the reduction of saturation charge to 50% of that in normal operation.

2.5.4 Multi-phase electrode/bit operation

A system of charge transfer which makes better use of chip area and as such has advantages for large-scale digital storage is the electrode/bit scheme. Each electrode has built in directionality as shown in Fig. 2.12 and has a separately identifiable phase. The phases are applied sequentially along the device (driven from a ring counter for example) so that n electrodes are capable of storing $n - 1$ bits of information as shown in Fig. 2.17. It is convenient to break into the cycle when all phase voltages from 1 to $n - 1$ are applied, allowing storage of their charge. ϕ_n is then applied but is storing a zero. The next event is to reduce ϕ_{n-1} to zero so that the charge under its electrode is transferred to ϕ_n electrode and the zero moves under ϕ_{n-1}. ϕ_{n-1} is reapplied and ϕ_{n-2} is removed so that the next bit is moved one cell along the CCD and the zero moves one further cell to the left. This procedure is repeated sequentially for each electrode or phase until the zero leaves the CCD from the left-hand side. All the information has thereby been moved one cell to the right. The entire sequence repeats periodically to make transfers completely through the CCD. Clearly, the clock circuit complexity is

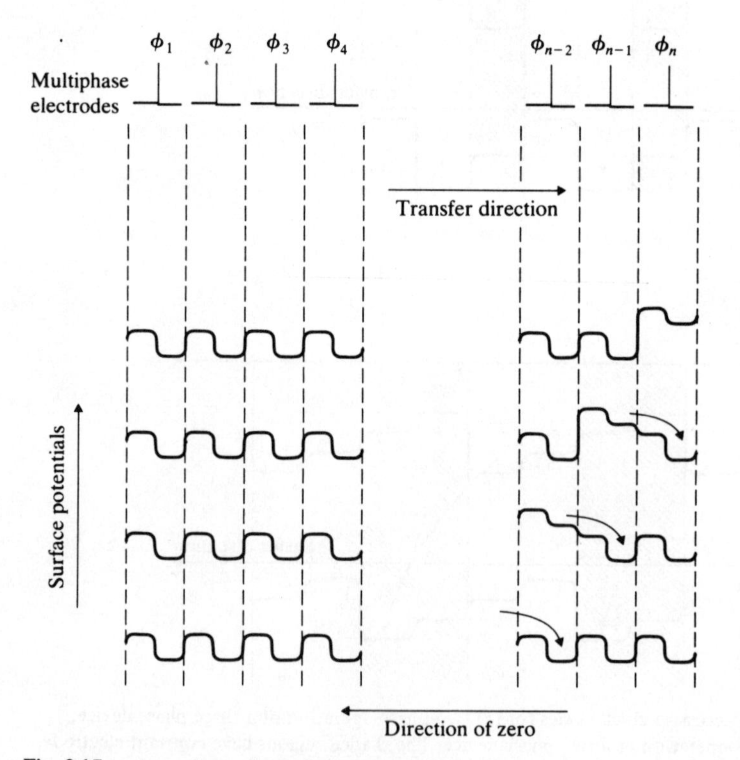

Fig. 2.17 An n-phase electrode/bit clock sequence.

considerable, but some simplifications will appear when the technique is applied to two-dimensional storage arrays.

2.6 The charge store in a buried channel

In later discussions of CCD defects it will be seen that containment of the signal charge in the bulk of the semiconductor may be preferable to containment at the surface. The potential minimum may be located away from the surface, as shown in Fig. 2.18, by introducing a thin layer of semiconductor with polarity opposite to that of the substrate. It is essential for the p–n junction in the buried channel device to be maintained in reverse bias so that the thin layer of n-type material would be fully depleted in the absence of desirable stored charge. This reverse bias is provided by a d.c. contact to the n-type region of Fig. 2.18, located at the output of the CCD. If the p-type and n-type materials are uniform the point of zero field (or potential minimum) occurs at a distance l from the edge of the depletion layer where, according to integration of Poisson's equation, we have

$$-\frac{l_1 e N_A}{\epsilon_S \epsilon_0} + \frac{(l - l_1) e N_D}{\epsilon_S \epsilon_0} = 0$$

i.e.
$$\frac{l - l_1}{l_1} = \frac{N_A}{N_D} \qquad 2.19$$

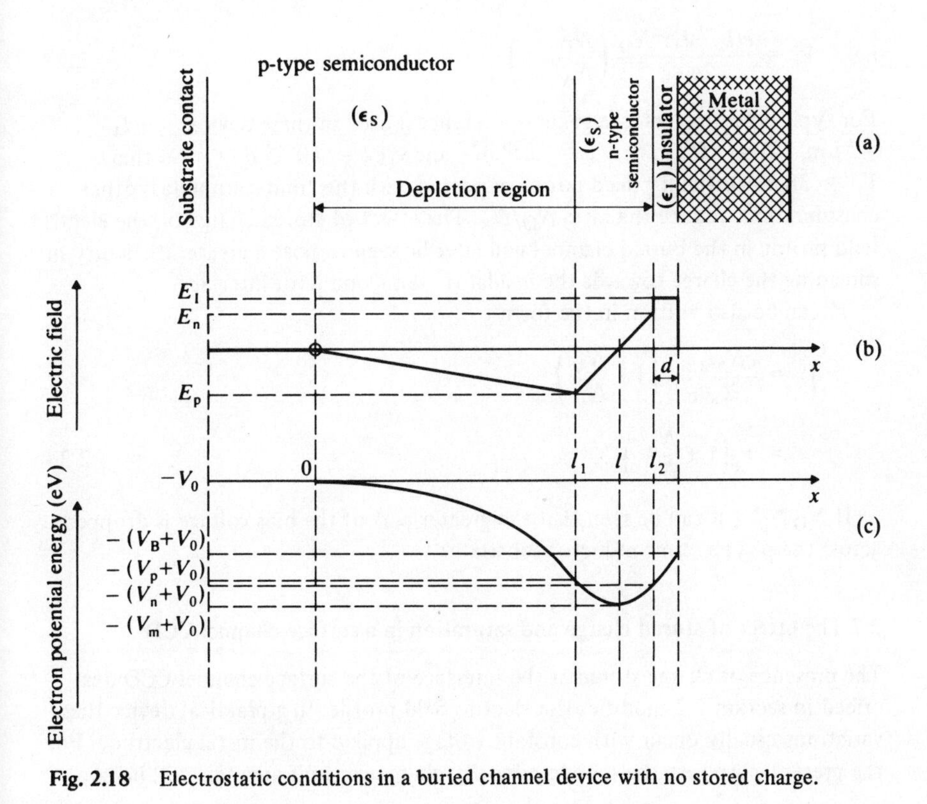

Fig. 2.18 Electrostatic conditions in a buried channel device with no stored charge.

N_A and N_D are the acceptor and donor densities in the p- and n-type semiconductors respectively. From Fig. 2.18, it can be seen that the buried channel device becomes a surface channel device when $l = l_2$. From equation 2.19 this limiting condition occurs when

$$l_1 = l_{1m} = (l_2 - l_1)N_D/N_A \tag{2.20}$$

The limiting voltage across the p-type substrate occurs when

$$V_p = \frac{eN_A l_{1m}^2}{2\epsilon_s\epsilon_o} = \frac{eN_D^2}{N_A}\frac{(l_2-l_1)^2}{2\epsilon_s\epsilon_o} \tag{2.21}$$

and the limiting voltage across the n-type layer occurs when

$$(V_n - V_p) = \frac{eN_D(l_2-l_1)^2}{2\epsilon_s\epsilon_o} \tag{2.22}$$

The insulator is usually much smaller in dimension d than the scale indicated in Fig. 2.18. Typically, it will only be 0.1 μm, with l_1 and $l_2 - l_1$ several micrometres deep. If we neglect the voltage drop across the insulator (care must be taken here because spurious surface charge could require the field in the insulator to be large) the transition from a buried transfer channel to a surface transfer channel (with no contained charge) will occur with an applied voltage, V_t, which equals V_n under the conditions of equations 2.20, 2.21 and 2.22.

$$\text{i.e.}\quad V_t = \frac{e(l_2-l_1)^2 N_D}{2\epsilon_s\epsilon_o}\left(\frac{N_D}{N_A}+1\right) \tag{2.23}$$

For typical parameters (see general references) used in these devices $l_2 - l_1 = 2\ \mu$m, $N_D = 10^{21}\ \text{m}^{-3}$, $N_A = 10^{20}\ \text{m}^{-3}$ and $\epsilon_s\epsilon_o = 10^{-10}\ \text{F m}^{-1}$. So that $V_t \simeq 38$ V. Difficulty need not be expected with this limitation unless other considerations reduce the ratio N_D/N_A. The effect of stored charge on the electric field profile in the buried channel will later be seen to pose a greater difficulty in spreading the charge towards the insulator–semiconductor interface.

V_t can be also written in the form

$$V_t = \frac{eN_A l_{1m}^2}{2\epsilon_s\epsilon_o}\left(1+\frac{N_A}{N_D}\right)$$

$$= V_p\left(1+\frac{N_A}{N_D}\right) \tag{2.24}$$

If $N_D \gg N_A$ it can be seen that the greater part of the bias voltage is dropped across the p-type semiconductor substrate.

2.7 The effect of stored charge and saturation in a surface channel CCD

The presence of charge stored at the interface of the surface channel CCD described in section 2.2 modifies the electric field profile. In a practical device these variations usually occur with constant voltage applied to the metal electrode. For the present purposes, the mobile minority charge per unit area, ρ_n, will be considered to reside in a sheet of zero thickness at the insulator–semiconductor interface. Estimates of the 'thickness' of this charge distribution will be made later. The modified electric field distribution is illustrated in Fig. 2.19. The depletion layer width has been reduced as dictated quantitatively by equation 2.12, and the potential difference in the insulator has increased. Using equation

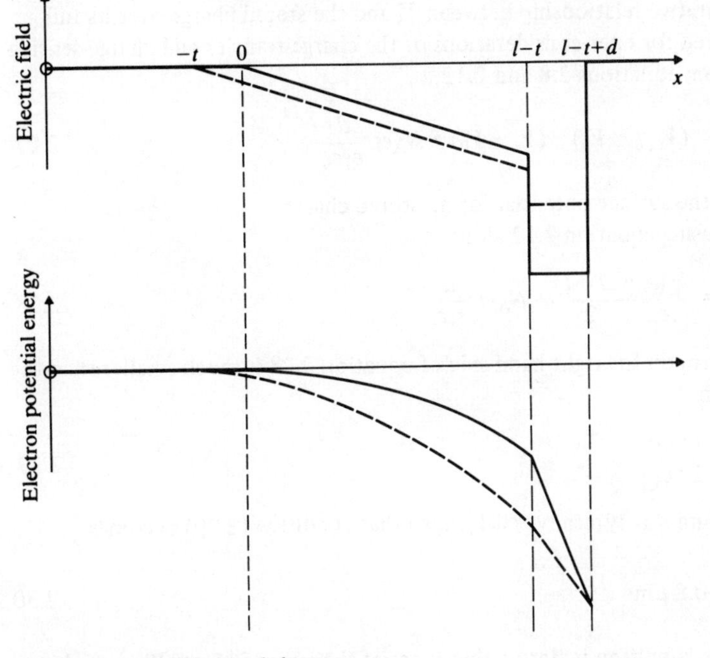

Fig. 2.19 Modifications caused by stored charge to the electrostatic conditions of a surface channel CCD.

2.12 or equation 2.17 the depletion layer will be completely removed when $\rho_n = \rho_{n,\text{sat}}$ given by

$$-\frac{\rho_{n,\text{sat}}d}{\epsilon_I\epsilon_0} = V_B - V_{FB} = \frac{N_A el}{\epsilon_I\epsilon_0}\left(\frac{\epsilon_I}{\epsilon_s}\frac{l}{2} + d\right) \tag{2.25}$$

providing that the fixed charge is not modified by the changes of potential energy at the interface. d is the insulator width.

Equation 2.25 may be written as

$$-\rho_{n,\text{sat}} = \frac{\epsilon_I\epsilon_0}{d}(V_B - V_{FB}) = N_A el\left(\frac{\epsilon_I}{\epsilon_s}\frac{l}{2d} + 1\right) \tag{2.26}$$

$\rho_{n,\text{sat}}$ provides a crude estimate of the maximum amount of charge that can be stored. It will be slightly optimistic because a significant inversion condition ($\simeq$1 V) will have to be maintained at the interface to avoid recombination of the stored minority electrons with the majority holes. It is interesting to note that equation 2.26 implies that the amount of charge stored can be greater than the amount of depleted charge, $N_A el$, for the empty potential well. For typical operating conditions at a silicon–silicon dioxide interface with $d = 0.125\ \mu$m and $\epsilon_I/\epsilon_s \simeq 0.25$, $\rho_{n,sat}$ will be twice the depleted charge for $l = 1\ \mu$m and will be ten times the depleted charge for $l = 9\ \mu$m. Also in equation 2.26, $\epsilon_I\epsilon_0/d$ is the oxide capacity per unit area, C_{ox}. It simply relates $\rho_{n,sat}$ to V_B.

The quantitative relationship between V_s and the stored charge over its full range is required for later considerations of the charge-transfer and charge-sensing processes. From equations 2.8 and 2.12,

$$-\frac{\rho_n d}{\epsilon_I\epsilon_0} = (V_{s,0} - V_0) - (V_s - V_0) + N_A et\,\frac{d}{\epsilon_I\epsilon_0} \tag{2.27}$$

where $V_{s,0}$ is the surface potential for no stored charge.

Similarly, using equation 2.12 alone,

$$-\frac{\rho_n d}{\epsilon_I\epsilon_0} = \frac{N_A e}{\epsilon_s\epsilon_0}\,\frac{t(2l - t)}{2} + N_A et\,\frac{d}{\epsilon_I\epsilon_0} \tag{2.28}$$

The second term on the right-hand side of equation 2.28 is much smaller than the preceding term if

$$d \ll \left(l - \frac{t}{2}\right)\frac{\epsilon_I}{\epsilon_s} \tag{2.29}$$

$\epsilon_I/\epsilon_s \simeq 0.25$ and d is typically $\simeq 0.1\ \mu$m so that condition (2.29) becomes

$$l - \frac{t}{2} \gg 0.8\ \mu\text{m} \tag{2.30}$$

In essence this condition is stating that most of the applied bias voltage of the storage cell with no contained charge must exist across the depletion layer. When this condition is satisfied, equation 2.27 shows that there is an approximately linear relationship between ρ_n and the surface potential. The same conclusion may be reached by inspection of equation 2.18,

$$\text{i.e.}\quad -\rho_n \simeq -(V_s - V_{s,0})(\epsilon_I\epsilon_0/d) = -(V_s - V_{s,0})\,C_{ox} \tag{2.31}$$

The small non-linear component of the surface potential in equation 2.27 is simply the change in depleted charge divided by the oxide capacity. Linearity requires that the change of depleted charge must be much less than the stored charge causing it.

For a typical CCD carrying a maximum charge per cell of 1 pC under an electrode of dimensions 10 μm x 100 μm on a silicon dioxide insulator with $d = 0.1\ \mu$m and $\epsilon_I = 2.5$, $(V_{s,0} - V_s)$ is approximately 4 V. A potential of at least this amount must be applied to the electrode (neglecting built in potentials due to fixed interface charge) for the CCD to carry 1 pC of saturation charge.

An alternative description of the above properties may be given in terms of the capacities per unit area of the insulator, C_{ox} ($= \epsilon_I\epsilon_0/d$), and of the depletion layer, C_{dep}. C_{dep} can be defined, with the aid of equation 2.8, as the ratio of the depleted charge, $N_A el$, and the depletion layer voltage or surface potential, $(V_s - V_0)$.

i.e. $$C_{dep} = 2\epsilon_s\epsilon_0/l \qquad 2.32$$

Care must be taken in using equation 2.32, because the ratio of a small increment of depleted charge to the corresponding change of surface potential is

$$C_{\Delta,\,dep} = \epsilon_s\epsilon_0/l \qquad 2.33$$

These difficulties are caused by the non-linear character of C_{dep} caused by the dependence of l on the surface potential.

Substituting equation 2.32 in equation 2.12 we have

$$V_B = N_A el\left(\frac{1}{C_{dep}} + \frac{1}{C_{ox}}\right) - \frac{\rho}{C_{ox}} \qquad 2.34$$

and using equation 2.8,

$$V_B = (V_s - V_0)\left(1 + \frac{C_{dep}}{C_{ox}}\right) - \frac{\rho}{C_{ox}} \qquad 2.35$$

$C_{dep}/C_{ox} \ll 1$ if $l/\epsilon_s \gg d/\epsilon_I$. Therefore equation 2.35 shows that there is an approximately linear relationship between ρ and the surface potential $(V_s - V_0)$. C_{ox} is the constant of proportionality.

2.8 Design considerations for a surface channel charge store

The charge store must be designed to give as large a dynamic range as possible, to avoid electrical breakdown and to be compatible with convenient operating voltages. At the small signal extreme the dynamic range is limited by noise and interference considerations. It can be maximized by making the saturation or full-well charge, $\rho_{n,\,sat}$, as large as possible. Under saturation conditions when $l = 0$, equation 2.6 shows that

$$\rho = \rho_{n,\,sat} + \rho_F = \epsilon_I\epsilon_0 E_I \qquad 2.36$$

A large $\rho_{n,\,sat}$ clearly requires E_I to be as large as possible. The ultimate limit will be set by dielectric breakdown conditions in the insulator. For good quality SiO_2, the limit is 900 MV m^{-1}, but 100 MV m^{-1} is probably a better limit to set for a practical device. With this latter figure the maximum sum of mobile and fixed interface charge is approximately -0.002 C m^{-2} or -2 pC on an electrode area of 10 μm by 100 μm. ρ_F for a p-type silicon substrate and SiO_2 insulator is typically $+0.001$ Cm^{-2}. The voltage across the insulator is $E_I d$ so that a bias voltage of at least this value is required to support the charge storage. The smallest value of d is set by making the SiO_2 thickness as small as technologically possible while avoiding 'pinhole' and related leakage effects. A thickness of 0.1 μm is typical. With $E_I = 100$ MV m^{-1} the required bias voltage is 10 V. If d were significantly larger than 0.1 μm the required bias voltage would be inconveniently high.

Having decided the saturation design conditions it is necessary to identify problems that may occur for smaller ρ_n. Substituting equation 2.14 in equation 2.17 we have

$$(V_B - V_{FB}) = \frac{\epsilon_s\epsilon_0}{2N_A e}E_s^{\,2} - \frac{\epsilon_s}{\epsilon_I}E_s d - \frac{\rho_n d}{\epsilon_I\epsilon_0} \qquad 2.37$$

Remembering that ρ_n is negative, E_s must change from zero to an increasing negative value when ρ_n is reduced from $\rho_{n,sat}$. The maximum electric field in the semiconductor occurs when $\rho_n = 0$. In this empty well condition E_s must be less than the breakdown field in the semiconductor which is approximately 30 MV m^{-1} for good quality silicon, but a practical limit may be 10 MV m^{-1}. The lower breakdown field in the semiconductor clearly requires that the bias voltage for zero stored charge must be spread over a depletion layer which is much deeper than the insulator. Equation 2.37 shows that this requirement may also be met by a small N_A. Coincidentally this is also the condition for good linearity in the relationship of surface potential and stored charge; a feature which will be seen to be valuable in later considerations of charge-sensing techniques. V_{FB} is typically −3 V for a p-type silicon substrate with SiO_2 insulator, so that zero stored charge exists with a positive electric field in the insulator and an electric field in the semiconductor which is larger than it would be if there was no interface charge. A lower limit to N_A set by the available technology is typically 10^{20} m^{-3}. To a good approximation in equation 2.37 we can neglect the voltage, $(\epsilon_s/\epsilon_I)E_s d$, dropped across the insulator when $\rho_n = 0$ so that the peak field in the semiconductor under these conditions is

$$E_{s,\,max} = \sqrt{\frac{2N_A e}{\epsilon_s\epsilon_0}(V_B - V_{FB})} \qquad 2.38$$

Taking $V_B - V_{FB} = 15$ V, $E_{s,\,max} \simeq 2.3$ MV m^{-1} for $N_A = 10^{20}$ m^{-3} and $E_{s,\,max} \simeq 7$ MV m^{-1} for $N_A = 10^{21}$ m^{-3}. The latter value is approaching the breakdown limit described above so that the substrate doping must typically lie between $N_A = 10^{20}$ m^{-3} and $N_A = 10^{21}$ m^{-3}.

2.9 The effect of stored charge and saturation in a buried channel CCD

In Fig. 2.18 charge would be stored at position l, providing the space-charge distortion of the electric field were insignificant. When the charge is sufficiently large there is a distortion of the electric field as shown in Fig. 2.20. The electric field is zero (neglecting diffusion) in the region of charge storage so that the stored-charge density is equal to the donor density. If the charge density were smaller there would be an electric field gradient causing the electric field near the edges of the space charge to 'point' inwards so that further bunching would occur. Correspondingly a stored electron density greater than the donor density would be reduced by outward pointing electric fields. If the charge were stored with invariant bias voltage, V_B, the two hatched areas in Fig. 2.20 would be equal. For analysis, it is convenient to consider the potential energy to the left of the space charge separately from that to the right. The effect of fixed charge at the semiconductor–insulator interface will be neglected. This may cause a bias-

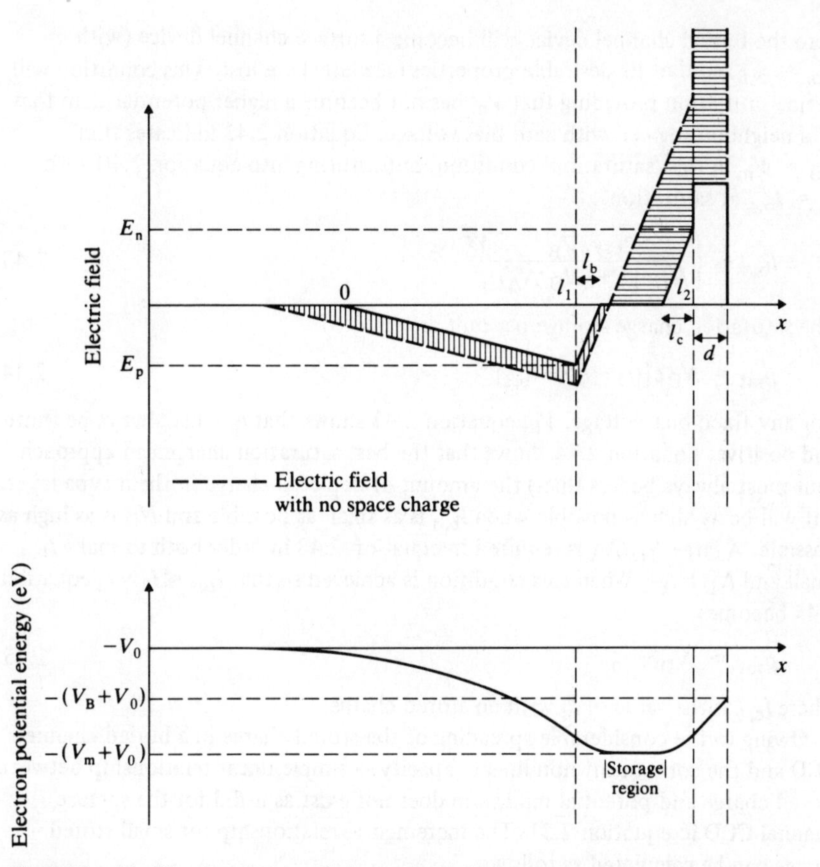

Fig. 2.20 Modifications by stored charge to the electrostatic conditions of a buried channel CCD.

voltage offset of several volts ($\rho_F d/\epsilon_I \epsilon_0$) in a practical device. On the left-hand side

$$V_m = -E_p\left(\frac{l_1}{2} + \frac{l_b}{2}\right) = -\frac{E_p l_b}{2}\left(\frac{N_D}{N_A} + 1\right) \qquad 2.39$$

$$\therefore \quad V_m = \frac{N_D e}{2\epsilon_s \epsilon_0} l_b^2\left(1 + \frac{N_D}{N_A}\right) \qquad 2.40$$

On the right-hand side,

$$(V_B - V_m) = -\frac{E_n l_c}{2} - \frac{\epsilon_s}{\epsilon_I} E_n d \qquad 2.41$$

$$= -\frac{N_D e}{2\epsilon_s \epsilon_0}\left(l_c^2 + 2l_c \frac{\epsilon_s}{\epsilon_I} d\right) \qquad 2.42$$

Initially the saturation properties will be considered. They are more complicated than in the case of the surface channel CCD. When l_c, in Fig. 2.20, becomes zero the buried channel device will become a surface channel device (with $V_m = V_B$) so that its desirable properties (see later) are lost. This condition will define saturation providing that V_m has not become a higher potential than that of a neighbouring cell with zero bias voltage. Equation 2.42 indicates that $V_B = V_m$ in this 'saturation' condition. Substituting into equation 2.40 with $l_b = l_{b,s}$ at saturation

$$l_{b,s} = \left\{\frac{2\epsilon_s\epsilon_0 V_B}{N_D e[1 + (N_D/N_A)]}\right\}^{\frac{1}{2}} \quad 2.43$$

The saturation charge storage per unit area is

$$\rho_{sat} = N_D e[(l_2 - l_1) - l_{b,s}] \quad 2.44$$

For any fixed bias voltage, V_B, equation 2.43 shows that l_b must always be finite and positive. Equation 2.44 shows that the best saturation charge can approach (but must always be less than) the amount of depleted charge in the n-type layer, but will be as high as possible when $l_{b,s}$ is as small as possible and N_D is as high as possible. A large N_D/N_A is required in equation 2.43 in order both to make $l_{b,s}$ small and N_D large. When this condition is achieved so that $l_{b,s} \ll l_2 - l_1$ equation 2.44 becomes

$$\rho_{sat} \simeq N_D e l_{c,0} \quad 2.45$$

where $l_{c,0}$ is the value of l_c with no stored charge.

Owing to the considerable spreading of the stored charge in a buried channel CCD and the consequent non-linear capacity, a simple linear relationship between stored charge and potential minimum does not exist as it did for the surface channel CCD in equation 2.31. The incremental relationship for small stored charge can be calculated as follows.

From equation 2.40,

$$\Delta l_b = \frac{\epsilon_s\epsilon_0 \Delta V_m}{N_D e l_{b,0}[1 + (N_D/N_A)]} \quad 2.46$$

$l_{b,0}$ is the value of l_b with no stored charge.

From equation 2.42

$$\Delta l_c = \frac{\epsilon_s\epsilon_0 \Delta V_m}{N_E e[l_{c,0} + (\epsilon_s/\epsilon_I)d]} \quad 2.47$$

$l_{c,0}$ is the value of l_c with no stored charge.

The stored charge per unit area is

$$\rho_n = N_D e(\Delta l_b + \Delta l_c)$$

$$\simeq \epsilon_s\epsilon_0 \left\{\frac{1}{l_{b,0}[1 + (N_D/N_A)]} + \frac{1}{[l_{c,0} + (\epsilon_s/\epsilon_I)d]}\right\} \Delta V_m \quad 2.48$$

Recognizing that $l_{b,0}N_D/N_A$ is equal to the depletion layer width in the substrate (equation 2.19) and is much larger than $l_{c,0}$ and also $l_{c,0} \gg (\epsilon_s/\epsilon_I)d$, equation 2.48 becomes

$$\rho_n \simeq \frac{\epsilon_s\epsilon_0}{l_{c,0}}\Delta V_m \qquad 2.49$$

The error of this linear approximation when extrapolated to saturation (where $\Delta V_m \simeq (N_D e/2\epsilon_s\epsilon_0)l_{c,0}^2$ from equation 2.42), is approximately a factor of 2 when compared with equation 2.45 where $\rho_{sat} \simeq N_D e l_{c,0}$. A typical variation of stored charge with the minimum potential energy is illustrated in Fig. 2.21.

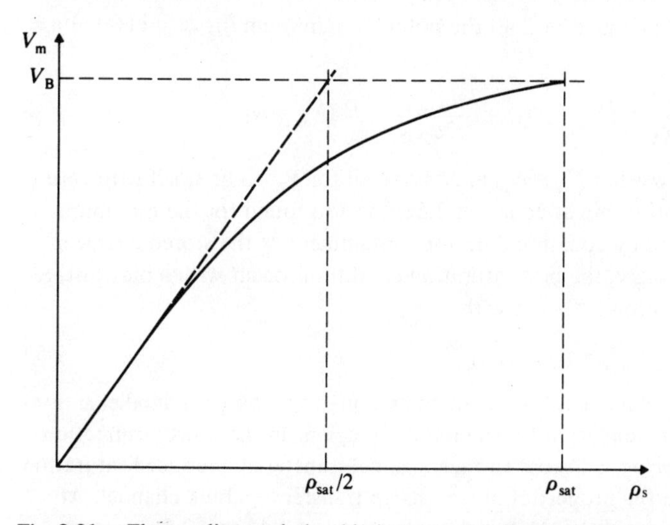

Fig. 2.21 The non-linear relationship between stored charge and the minimum potential energy in a buried channel CCD with uniform doping in the channel layer.

2.10 Design considerations for a buried channel charge store

Equation 2.43 indicates that V_B should be as small as possible for good saturation properties. However it must not be smaller than the depth of a neighbouring potential minimum containing no charge and having no applied bias voltage. Under these conditions we can define

$$l_b = l_{b,0,0} \qquad 2.50$$

and $$l_c = (l_2 - l_1) - l_{b,0,0} \qquad 2.51$$

From equations 2.40 and 2.42 with $V_B = 0$,

$$\frac{N_D e}{2\epsilon_s\epsilon_0} l_{b,0,0}^2\left(1 + \frac{N_D}{N_A}\right) = \frac{N_D e}{2\epsilon_s\epsilon_0}\left\{[(l_2 - l_1) - l_{b,0,0}^2] + 2[(l_2 - l_1) - l_{b,0,0}]\frac{\epsilon_s}{\epsilon_I}d\right\}$$

i.e. $$\frac{N_D}{N_A} l_{b,0,0}^2 + 2l_{b,0,0}\left[(l_2 - l_1) + \frac{\epsilon_s}{\epsilon_I}d\right] - (l_2 - l_1)\left[(l_2 - l_1) + \frac{2\epsilon_s}{\epsilon_I}d\right] = 0 \qquad 2.52$$

If we make the following assumptions that will be marginally met in practice,

$$2(\epsilon_s/\epsilon_I)d \ll (l_2 - l_1), \tag{2.53}$$

$$N_D \gg N_A \tag{2.54}$$

we have from equation 2.51

$$l_{b,0,0}^2 \simeq (N_A/N_D)(l_2 - l_1)^2 \tag{2.55}$$

We have already seen the need in equation 2.43 for N_A/N_D to be small. Substituting equation 2.55 in equation 2.40 the potential minimum for zero bias voltage is

$$V_m = \frac{e(l_2 - l_1)^2}{2\epsilon_s\epsilon_0}(N_A + N_D) = \frac{e l_{b,0,0}^2}{2\epsilon_s\epsilon_0}\frac{N_D}{N_A}(N_A + N_D) \tag{2.56}$$

Equation 2.56 shows that $N_A + N_D$ must be small for V_m to be small with zero bias. The same relationship as equation 2.56 was also found for the minimum bias voltage required by equation 2.43 for containment of the stored charge at saturation. Accordingly, the best saturation conditions occur with a bias voltage equal to V_m in equation 2.56 and with

$$l_{b,0,0} = l_{b,s} = (l_2 - l_1)(N_A/N_D)^{\frac{1}{2}} \tag{2.57}$$

A smaller bias voltage than that defined by equation 2.56 gives smaller saturation charge owing to spillage into neighbouring regions in the transfer direction. A larger bias voltage would cause surface channel transfer at a reduced saturation charge and degrade the properties of the charge transfer in a bulk channel. At first sight the equality of the relationship of equations 2.43 and 2.56 appears surprising but it may be clarified by the spatial distribution of the electric field for zero bias, zero stored charge with optimum bias and saturation charge with optimum bias. These distributions are illustrated in Fig. 2.22. If we neglect the voltage across the insulator the areas on either side of the potential minimum (zero field) in the first case must be equal and each must be equal to the optimum bias. In the second case the areas must differ by the optimum bias. In the third case the left-hand area must equal the optimum bias and the right-hand area is zero. In this way the potential energy minimum with no stored charge and no bias voltage is equal to the potential energy minimum with saturation stored charge and bias voltage applied.

To give a good quality p-type substrate N_A should not be less than approximately 10^{20} m^{-3}. Also $(l_2 - l_1)$ should not be less than about 2 μm to give a good epitaxial layer and circuit convenience requires the bias voltage (equal to V_m in equation 2.56) to be not much greater than 10 V. Using these values in equations 2.56 and 2.57

$$N_D + N_A \simeq 3.1 \times 10^{21}\ m^{-3}$$

i.e.
$$N_D \simeq 3 \times 10^{21}\ m^{-3} \tag{2.58}$$

and
$$l_{b,s}/(l_2 - l_1) \simeq 0.18 \tag{2.59}$$

The bias voltage in equation 2.56 corresponding to the optimum design condition in equation 2.57 would have to be offset by an amount equal to the extra insulator voltage, $\rho_F d/\epsilon_I \epsilon_0$, caused by spurious fixed charge at the semiconductor–insulator interface. The same offset would also have to be applied to the non-store condition with zero-bias voltage.

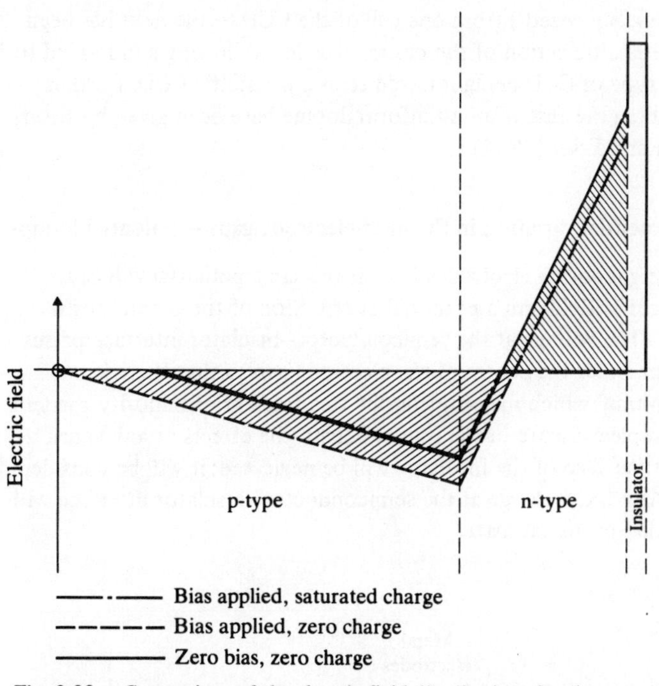

Fig. 2.22 Comparison of the electric field distributions for three conditions of bias voltage and stored charge in a buried channel CCD.

It is difficult to compare the saturation charge of a surface channel CCD and a buried channel CCD. The former can carry an amount of charge significantly larger than the amount of depleted charge (equation 2.26). The latter cannot carry as much charge as is depleted but the depletion is in a material with a significantly higher doping concentration (equations 2.44 and 2.58). Using the parameters relevant to equation 2.58 the saturation charge is 8.6×10^{-4} C m^{-2} or approximately 1 pC under an electrode with dimensions of 10 μm x 100 μm.

If N_D is made too large there is a danger of avalanche breakdown in the semiconductor. If $l_{b,0,0} \ll l_2 - l_1$ the highest electric field occurs at the semiconductor–insulator interface with zero bias voltage. A worst case estimate for E_{max} at this point can be obtained by assuming $l_{c,0,0} \simeq l_2 - l_1$.

i.e. $$E_{max} = N_A e(l_2 - l_1)/\epsilon_S \epsilon_0 \qquad 2.60$$

A similar limit will be reached at the interface of the p- and n-regions of the semiconductor at the limiting bias given by equation 2.23 when the device operates with surface channel transfer for small charge. Using the parameters relevant to equation 2.58,

$$E_{\max} = 10.35 \text{ MV m}^{-1}$$

This is close to the limit described in section 2.8.

The reduction in width of the stored space charge as it is transferred (in appearance it is being 'squeezed') from one cell of the CCD to the next has been compared to the peristaltic action of the oesophagus in swallowing and has led to one version of this type of CCD being referred to as a peristaltic CCD. Further design details including the case of non-uniform doping have been given by Esser, (1972), Theunissen and Esser (1974).

2.11 The surface potential variation in the inter-electrode gaps – potential humps

The gap between neighbouring electrodes held at the same potential will cause distortion of the electric field and a potential distribution of the general form shown in Fig. 2.23. The voltage at the semiconductor–insulator interface passes through a minimum under the centre of the electrode gap so that there is a surface-potential 'hump' which may impede the free transfer of minority carriers. In the following simple estimate of the hump 'height' the effects of any spurious charge on the upper surface of the insulator will be neglected; it will be considered in a later section. Any fixed charge at the semiconductor–insulator interface will be implicitly included in the estimate.

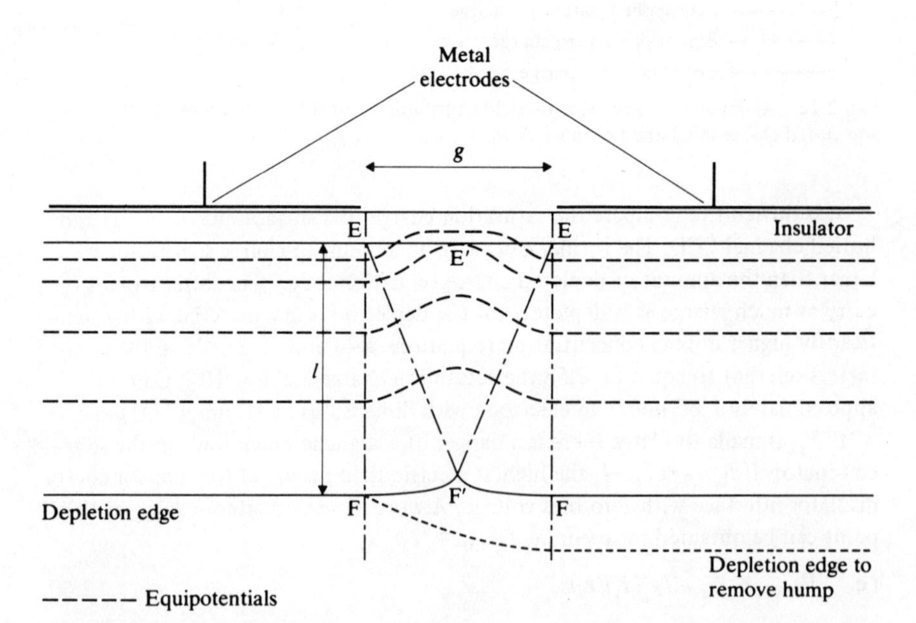

Fig. 2.23 Potential humps arising from distortion of the electrostatic conditions in the gap between neighbouring electrodes.

The depletion of majority carriers is essentially caused by repulsion of the positive holes by the positive charge, which the bias places on the metal electrode, to a distance where the electrostatic force is screened by the negative charge of the acceptors in the depletion layer. In a simple uniform geometry where the electrode extends to infinity the depletion edge is the same distance from the metal electrode at all points. It will be assumed that the depletion edge is a constant distance from the metal electrodes even in the region under the electrode gaps, as shown in Fig. 2.23 by the circular arcs FF′. This approximation neglects the additional screening effect of acceptors in the depletion region under the electrode gap but it will become asymptotically true when $l \gg g/2$. We will further assume that the electrode gap, g, is small enough for complete depletion to occur even at the centre of the gap at E′. Along the line E′F′ and along a similar line perpendicular to the centre of the electrode, symmetry requires the divergence of the electric field to be perpendicular to the surface so that the variation of E with distance relative to the edge of the depletion layer is given by equation 2.3. Therefore the surface potential at E′, $V_{SH} - V_0$, is given by (equation 2.8):

$$V_{SH} - V_0 = \frac{N_A e[l^2 - (g^2/4)]}{2\epsilon_s\epsilon_0} \qquad 2.61$$

The surface potential underneath the electrode is

$$V_s - V_0 = \frac{N_A e l^2}{2\epsilon_s\epsilon_0} \qquad 2.62$$

The potential hump height in electron-volts, V_H, is then given by

$$V_H = -(V_{SH} - V_s) = \frac{N_A e g^2}{8\epsilon_s\epsilon_0} \qquad 2.63$$

In these equations it has been assumed that the centre of the circular arcs FF′ lies at E. It will lie at some point between E and the metal electrode but for any practical device with an insulator thickness of 0.1 μm and a depletion depth of several micrometres the error will be no worse than that caused by the other approximations.

V_H was calculated for equal voltages on neighbouring electrodes. Another important parameter for later considerations of charge transfer is the difference of electrode voltage necessary to remove the potential hump. For this estimate a higher bias voltage will be applied to the right-hand electrode in Fig. 2.23. There will always be a hump if any part of the depletion edge is closer than the distance l to the semiconductor–insulator interface. Using the same approximations as in the estimate of the hump height the hump will just be removed when the depletion layer under the right-hand electrode has become deep enough for its continuation, as a circular arc from G, to intersect the depletion edge under the left-hand electrode at F. The voltage difference, V_w, required to achieve this is simply the difference of surface potentials under the neighbouring electrodes.

i.e. $$V_{\mathrm{W}} = \frac{N_{\mathrm{A}} e g^2}{2\epsilon_s \epsilon_0} \qquad 2.64$$

In equations 2.63 and 2.64 it can be seen that V_{H} and V_{w} are independent of bias voltage. Numerical estimates for $g = 2\ \mu\mathrm{m}$ relevant to aluminium gate electrodes are

$$\begin{aligned} N_{\mathrm{A}} &= 10^{20}\ \mathrm{m}^{-3}; \quad V_{\mathrm{H}} = 0.09\ \mathrm{V} \text{ and } V_{\mathrm{w}} = 0.36\ \mathrm{V} \\ N_{\mathrm{A}} &= 5 \times 10^{20}\ \mathrm{m}^{-3}; \quad V_{\mathrm{H}} = 0.45\ \mathrm{V} \text{ and } V_{\mathrm{w}} = 1.8\ \mathrm{V} \\ N_{\mathrm{A}} &= 2 \times 10^{21}\ \mathrm{m}^{-3}; \quad V_{\mathrm{H}} = 1.8\ \mathrm{V} \text{ and } V_{\mathrm{w}} = 7.2\ \mathrm{V} \end{aligned} \qquad 2.65$$

For $g = 0.1\ \mu\mathrm{m}$, relevant to the polysilicon gate devices described later, the corresponding estimates are

$$\begin{aligned} N_{\mathrm{A}} &= 10^{20}\ \mathrm{m}^{-3}; \quad V_{\mathrm{H}} = 0.22\ \mathrm{mV} \text{ and } V_{\mathrm{w}} = 0.88\ \mathrm{mV} \\ N_{\mathrm{A}} &= 5 \times 10^{20}\ \mathrm{m}^{-3}; \quad V_{\mathrm{H}} = 1.1\ \mathrm{mV} \text{ and } V_{\mathrm{w}} = 4.4\ \mathrm{mV} \\ N_{\mathrm{A}} &= 2 \times 10^{21}\ \mathrm{m}^{-3}; \quad V_{\mathrm{H}} = 4.3\ \mathrm{mV} \text{ and } V_{\mathrm{w}} = 17.2\ \mathrm{mV} \end{aligned} \qquad 2.66$$

If more accurate estimates are required Poisson's equation must be solved for the particular geometry under consideration. Numerical techniques are required and have been developed by Amelio (1972), McKenna and Schryer (1973) and Collet and Vliegenthart (1974). The above estimates are satisfactorily accurate for many applications.

2.12 The charge-transfer rate

2.12.1 The initial transfer process

When the potential wall containing charge in a particular well is removed by application of a bias voltage to the following well, charge transfer will commence and will continue until the bias has been removed from the first well. If there is a uniform distribution of charge initially, the minority carrier repulsion will cause carrier motion along the interface (or potential valley in a buried-channel device). This process will only be effective over a short distance comparable with the insulator thickness d, near to the edge of the electrode owing to the attractive force of charge of opposite polarity on the electrode. The order of magnitude of the time required to remove this quantity of edge charge can be estimated from the repulsive force between two sections of charge of length d, in the transfer direction and separated by d, as illustrated in Fig. 2.24. If the charge per unit area is ρ_{n}, the charge on a section of length d and unit width of channel is $\rho_{\mathrm{n}} d$. The electric field at a distance d from this charge (in cylindrical geometry) is E_{d}, given by

$$E_{\mathrm{d}} = \frac{\rho_{\mathrm{n}} d}{2\pi\epsilon_s \epsilon_0 d} \qquad 2.67$$

If the mobility of the minority carriers is μ the induced drift velocity of those carriers at the edge of the electrode is μE_{d} and the characteristic time to remove a section of charge of length d is τ_R given by

$$\tau_R = \frac{d}{\mu E_d} = \frac{2\pi\epsilon_s\epsilon_0 d}{\mu\rho_n} \qquad 2.68$$

If $\rho_{n,\,sat}$ is the saturation charge per unit area equation 2.68 may be written as

$$\tau_R = \frac{2\pi\epsilon_s\epsilon_0 d}{\mu\rho_{n,\,sat}} \frac{\rho_{n,\,sat}}{\rho_n} \qquad 2.69$$

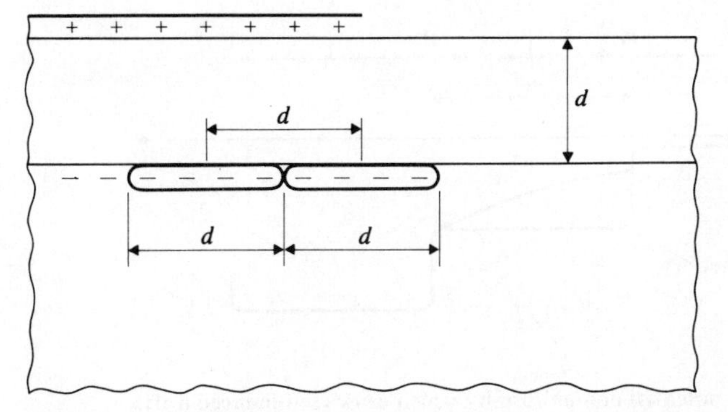

Fig. 2.24 The charge involved in mutual repulsion at the edge of an electrode.

The mobility of electrons in the bulk of silicon would be $\sim 0.1\ m^2\,V^{-1}\,s^{-1}$ at room temperature, but in the interface region a value of $0.05\ m^2\,V^{-1}\,s^{-1}$ is more typical. If $d = 0.1\ \mu m$ and $\rho_{n,\,sat} = 10^{-2}\ Cm^{-2}$ equation 2.69 shows that

$$\tau_R \simeq 10^{-12}\,s \quad \text{for} \quad \rho_n = \rho_{n,\,sat}$$

$$\text{and} \quad \tau_R \simeq 10^{-10}\,s \quad \text{for} \quad \rho_n = \rho_{n,\,sat}/100 \qquad 2.70$$

The initial very rapid transfer will also be assisted by diffusion and by any slope of surface potential in the edge region which is caused by the potential difference of neighbouring electrodes. All of these effects will produce a non-uniformity of the stored charge which allows the following mechanisms of self-induced drift and diffusion to operate.

2.12.2 Self-induced drift

If the charge density under an electrode is non-uniform it causes a corresponding variation of surface potential as shown in Fig. 2.25 during the time that the bias is being removed from the centre electrode. As was seen in equation 2.31 the surface potential is linearly proportional to the contained charge. Even though this linearity is not so clearly evident in the case of the buried channel device (Fig. 2.21) the accuracy of the simple estimates presented here will not prevent them being applied equally well to this type of device. The slope of the surface potential in Fig. 2.25 provides a self-induced electric field in the charge-transfer direction which only occurs during the transfer process.

Assuming that saturation occurs when the surface potential has increased by the bias voltage V_B, the surface potential at any point under the central electrode of Fig. 2.25 is given by

$$\frac{V}{V_B} = \frac{\rho_n}{\rho_{n,\,sat}} \qquad 2.71$$

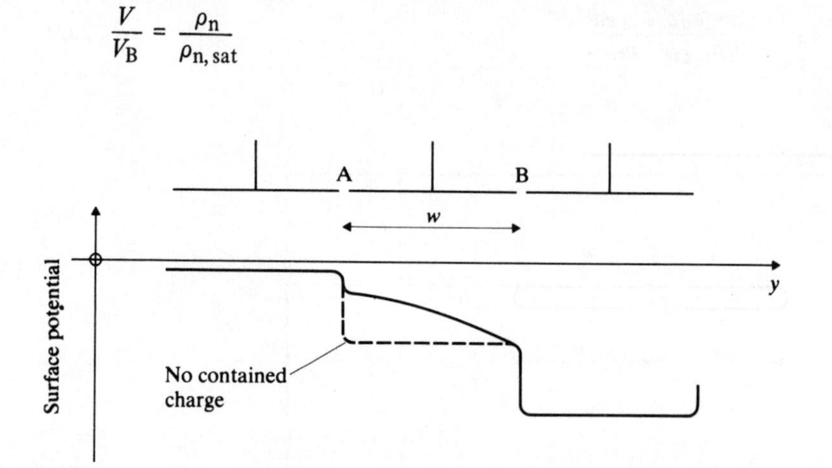

Fig. 2.25 The surface potential non-uniformity which causes self-induced drift.

ρ_n is a function of position, y, along the interface and the total stored charge is $\int_0^w \rho_n b\, dy$ where b is the channel width of the CCD. By considering the net flow of charge into any increment, dy, the rate of change of ρ_n at any point is given by

$$\frac{d\rho_n}{dt} = -\mu \frac{d}{dy}(\rho_n E) \qquad 2.72$$

$E = -\dfrac{dV}{dy} = -\dfrac{V_B}{\rho_{n,\,sat}}\dfrac{d\rho_n}{dy}$ using equation 2.71, so that equation 2.72 becomes

$$\frac{d\rho_n}{dt} = \frac{\mu V_B}{\rho_{n,\,sat}}\frac{d}{dy}\left(\rho_n \frac{d\rho_n}{dy}\right)$$

i.e.
$$\frac{d\rho_n}{dt} = \frac{\mu V_B}{2\rho_{n,\,sat}}\frac{d^2(\rho_n^{\,2})}{dy^2} \qquad 2.73$$

Equation 2.73 has the time and space variables on opposite sides of the equation so that the solution for ρ_n takes the form of a product of time and space functions. Neglecting mathematical rigour we can take a convenient solution for the spatial variation of ρ_n, with a similar shape to that shown in Fig. 2.25, in the form

$$\rho_n^2 = \rho_{n,p}^2 \cos\left(\frac{\pi y}{2w}\right) \qquad 2.74$$

$\rho_{n,p}$ is the value of ρ_n at the left-hand edge of the centre electrode in Fig. 2.25.
Substituting equation 2.74 in equation 2.73,

$$\frac{d\rho_n}{dt} = -\frac{\pi^2 \mu V_B}{8w^2 \rho_{n,sat}} \rho_n^2 \qquad 2.75$$

Integrating equation 2.75 with $\rho_n = \rho_{n,o}$ at time $t = 0$,

$$\frac{1}{\rho_n} - \frac{1}{\rho_{n,o}} = \frac{\pi^2 \mu V_B}{8w^2 \rho_{n,sat}} t$$

i.e.

$$\frac{\rho_n}{\rho_{n,o}} = \frac{1}{1 + \frac{\pi^2 \mu V_B}{8w^2} \frac{\rho_{n,o}}{\rho_{n,sat}} t} \qquad 2.76$$

From equation 2.76 we can identify a characteristic time, T_s, to describe the self-induced charge-transfer rate. T_s is the time required to reduce the charge to half its initial value.

i.e.

$$T_s = \frac{8w^2}{\pi^2 \mu V_B} \frac{\rho_{n,sat}}{\rho_{n,o}} \qquad 2.77$$

Even though the charge transfer is not a simple exponential decay process it is often convenient to describe its rate by the instantaneous time constant, τ_s, while recognizing that τ_s will vary with time. If the transfer of ρ_n is written as $\rho_n = \rho_{n,o} \exp(-t/\tau_s)$ we have

$$\tau_s = -\frac{1}{(1/\rho_n)(d\rho_n/dt)} \qquad 2.78$$

From equation 2.75,

$$\tau_s = \frac{8w^2}{\pi^2 \mu V_B} \frac{\rho_{n,sat}}{\rho_n} \qquad 2.79.$$

Comparison of equation 2.79 with equation 2.77 shows that $\tau_s = T_s$ in the initial stages of transfer by this process.

In making numerical estimates, $\mu V_B/w$ must not be allowed to exceed $\sim 10^5$ $m s^{-1}$ as this is approximately the maximum possible (saturated) drift velocity for electrons or holes in silicon. For $\mu = 0.05 \; m^2 V^{-1} s^{-1}$, $V_B = 10$ V and $w = 10 \; \mu m$,

$$\frac{8w^2}{\pi^2 \mu V_B} = 1.6 \times 10^{-10} \; s \qquad 2.80$$

Accordingly,

$$\tau_s = 1.6 \times 10^{-10} \; s \quad \text{for} \quad \rho_n = \rho_{n,sat}$$

and $\tau_s = 1.6 \times 10^{-8}$ s for $\rho_n = \rho_{n,sat}/100$

From equation 2.76

$$\rho_n \simeq \rho_{n,o}/100 \text{ when } t = 1.6 \times 10^{-8}\,\text{s} \quad \text{for} \quad \rho_{n,o} = \rho_{n,\text{sat}}$$
$$\text{and} \quad \rho_n \simeq \rho_{n,o}/2 \text{ when } t = 1.6 \times 10^{-8}\,\text{s} \quad \text{for} \quad \rho_{n,o} = \rho_{n,\text{sat}}/100 \tag{2.81}$$

From equation 2.79 it is clear that the process becomes successively slower as ρ_n becomes smaller. When $\rho_n \simeq \rho_{n,\text{sat}}/100$ we will see that the diffusion process becomes more effective as illustrated in Fig. 2.26. In Fig. 2.26 $\rho_{n,\text{sat}} = 10^{-3}$ C m^{-2} or 1 pC on an electrode with $w = 10\ \mu\text{m}$ and $b = 100\ \mu\text{m}$.

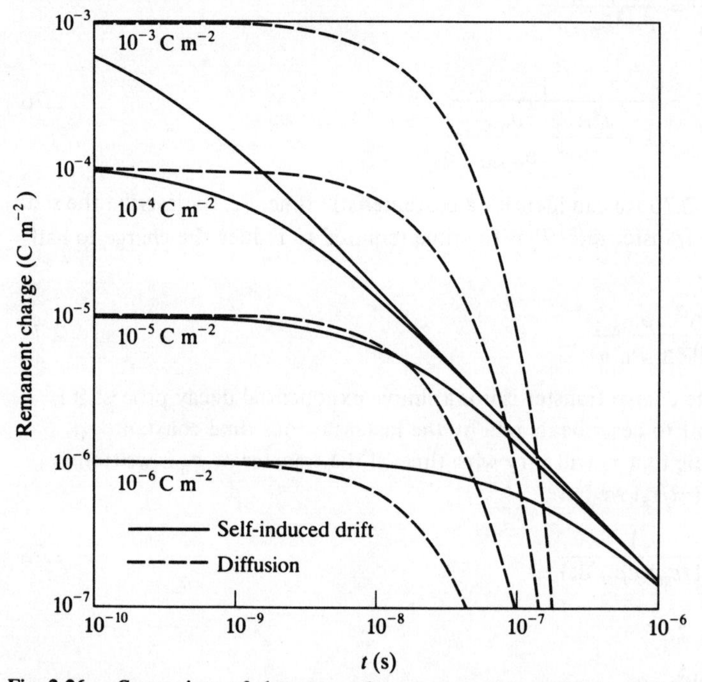

Fig. 2.26 Comparison of charge-transfer rates caused by self-induced drift and diffusion. The initial charge is shown as a parameter on the left-hand side. The saturation charge is 10^{-3} C m^{-2}.

Equation 2.76 and Fig. 2.26 show that the residual charge after a finite transfer time depends on the initial charge in a non-linear fashion. A similar problem is encountered in the bucket-brigade devices discussed in the next chapter where it is shown that it is necessary to have a minimum d.c. addition to any signal to avoid this non-linearity. This type of problem does not occur if the decay process is exponential because the residual charge is always a fixed proportion of the initial charge. Fortunately, in CCD the diffusion process which becomes important after the initial very rapid self-induced charge transfer, is an exponential process which helps to reduce the residual non-linearity. In a CCD this type of non-linearity is usually dominated by spurious trapping effects described later, or else the residual charge is so small that it is not troublesome. Further comments will be made when bucket-brigade devices are considered.

2.12.3 Diffusion

The charge gradient along the charge-transfer direction will also allow diffusion of charge. For simplicity the diffusion rate will be calculated for a linear charge gradient under the central electrode of Fig. 2.25 as shown in Fig. 2.27. If $\rho_{n,\,m}$ is the mean mobile charge per unit area and D is the diffusion coefficient

$$w\,\frac{d\rho_{n,\,m}}{dt} = D\,\frac{d\rho_n}{dy} \qquad 2.82$$

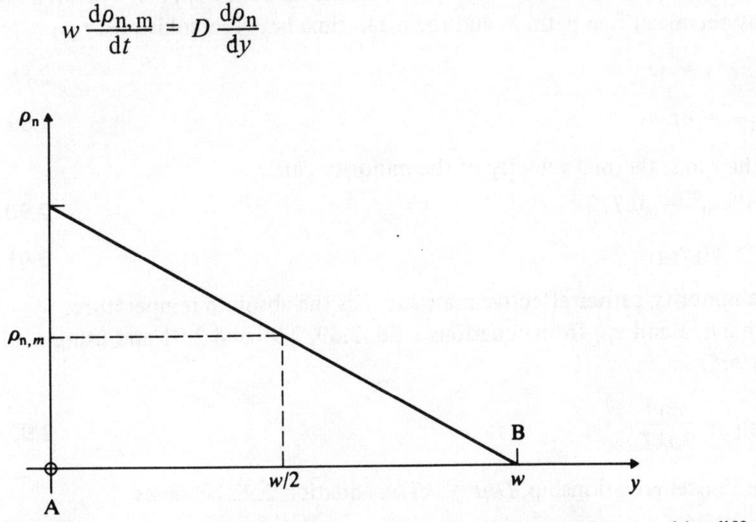

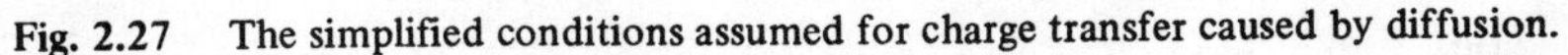

Fig. 2.27 The simplified conditions assumed for charge transfer caused by diffusion.

and $$\rho_n = 2\rho_{n,\,m}\left(\frac{w-y}{w}\right) \qquad 2.83$$

Substituting equation 2.83 in equation 2.82

$$\frac{d\rho_{n,\,m}}{dt} = -\frac{2D}{w^2}\,\rho_{n,\,m} \qquad 2.84$$

Integrating equation 2.84 with $\rho_{n,\,m} = \rho_{n,\,m,\,o}$ when $t = 0$

$$\rho_{n,\,m} = \rho_{n,\,m,\,o}\exp(-2Dt/w^2) \qquad 2.85$$

The time constant of this charge-transfer process is, τ_{Diff}, given by

$$\tau_{Diff} = \frac{w^2}{2D} \qquad 2.86$$

For $w = 10\ \mu m$ and $D = 25 \times 10^{-4}\ m^2\,s^{-1}$

$$\tau_{Diff} = 2 \times 10^{-8}\ s \qquad 2.87$$

Comparison of equation 2.87 with equation 2.81 shows that diffusion becomes more rapid than self-induced drift when $\rho_n \lesssim 10^{-2} \rho_{n,sat}$. For many applications the residual charge not transferred from an initial saturated charge must be less than one part in 10^4. This implies that the diffusion process must make the last hundredfold reduction which will require almost five time constants, which is 10^{-7} s using equation 2.87. This will give a maximum charge-transfer frequency of $10^7/2\pi = 1.6$ MHz.

A result similar to equation 2.86 can be estimated by equating τ_{Diff} to the time it takes a minority carrier to make a collision-controlled random walk along the distance w. The n collisions which occur in this random walk are separated on average by the mean free path, λ, and the mean time between collisions, τ_{CD}.

$$\therefore \quad \sqrt{(n\lambda^2)} = w \qquad 2.88$$

$$\text{and} \quad \tau_{Diff} = n\tau_{CD} \qquad 2.89$$

If v_{th} is the r.m.s. thermal velocity of the minority carriers,

$$\tfrac{1}{2}m^* v_{th}^2 = 3kT/2 \qquad 2.90$$

$$\text{and} \quad \lambda \simeq v_{th}\tau_{CD} \qquad 2.91$$

m^* is the minority carrier effective mass and T is the absolute temperature. Eliminating n, λ and v_{th} from equations 2.88, 2.89, 2.90 and 2.91 and using $\mu = e\tau_{CD}/m^*$,

$$\tau_{Diff} = \frac{ew^2}{3\mu kT} \qquad 2.92$$

Using the Einstein relationship, $D/\mu = kT/e$, equation 2.92 becomes

$$\tau_{Diff} = w^2/3D \qquad 2.93$$

The difference of the numerical factors in equations 2.86 and 2.93 is indicative of the accuracy of these estimates.

2.12.4 'Built-in' electric fields

The diffusion process takes about five time constants to reduce the saturation charge to the low residual considered above. This time is 10^{-7} s which may be too long when high-speed operation is desired. In this case it is necessary to aid the final stages of charge transfer with a built-in electric field which becomes dominant after the initial rapid transfer due to self-induced drift. If the built-in electric field is E_b the charge will drift from one potential minimum to the next in a time τ_b where

$$\tau_b = w/\mu E_b \qquad 2.94$$

With $\mu \simeq 0.1\ m^2V^{-1}s^{-1}$, in the bulk silicon, there is no point in making E_b greater than about $10^6\ Vm^{-1}$ owing to the onset of electron drift velocity saturation at $10^5\ m\ s^{-1}$.

In principle the built-in electric field could be produced by a modification of the stepped-oxide or diffused-implant two-phase CCD structures to give a surface potential slope in the transfer direction. However, there would be technological problems in achieving the desired lateral grading of the device properties. Instead, the built-in electric field is achieved with buried channel devices as illustrated in Fig. 2.28. For an isolated electrode on the insulator–semiconductor medium equipotentials close to the electrode are parallel to it except in the fringing regions so that electric fields parallel to the surface are zero under the centre of

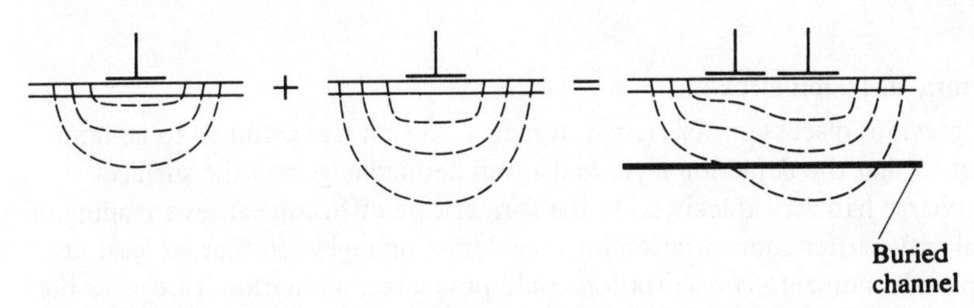

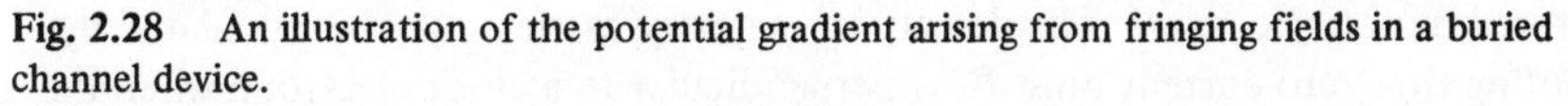

Fig. 2.28 An illustration of the potential gradient arising from fringing fields in a buried channel device.

the electrode. At lower potentials further into the semiconductor the equipotentials become curved so that the derivative of potential in the charge-transfer direction is significantly non-zero for a greater extent underneath the electrode. If a second electrode with a higher bias voltage is brought close to the first, its corresponding equipotentials lie deeper into the semiconductor. The combined effect causes the equipotential lines to deviate from parallelism with the electrode making the electric field in the charge-transfer direction non-zero, even under the centre of the electrode with the lower voltage. Two effects can now be recognized. The electric field in the charge-transfer direction will tend to increase with depth owing to the increasing angle between the equipotential lines and the electrode, but it will tend to decrease with depth owing to the greater spacing of equipotential lines at greater depth (see Fig. 2.1). Collet and Vliegenthart (1974) showed that the highest field under the centre of an electrode would occur at a depth of 0.4 of the electrode repeat distance. At this depth the electric field is typically about half the ratio of ΔV_B and the centre separation of the electrodes. For this reason a buried channel device designed to have its channel with zero charge at this depth will give high-speed charge transfer. For a large amount of stored charge when the containment channel is broad the self-induced fields cause rapid transfer as already described. The last fraction of charge will transfer at the depth of maximum field. τ_b could be as small as 100 ps for $w = 10\ \mu m$ and $\mu E_b = 10^5\ m\ s^{-1}$ (the saturated drift velocity). In this way devices have been clocked at rates in excess of 100 MHz (Theunissen and Esser, 1974).

2.12.5 Accurate calculations

All the above estimates have been much simplified, assumed independent and are necessarily inexact. In particular, the very rapid initial charge-transfer rates are much faster than the edges of the clock waveforms, so that the charge transfer is likely to be in quasi-static equilibrium with the rising surface potential of a well which is releasing its charge. Finite-difference computational methods are usually required in order to make quantitatively accurate calculations of all the effects as they simultaneously act (Fawcett and Vanstone, 1976). Other calculations using more accurate analytical techniques but giving the same order of magnitude estimates have been given by Carnes *et al* (1972), Amelio (1972) and Kim and Lenzlinger (1971).

2.13 Static diffusion effects

In the previous discussions of charge storage, diffusion was assumed to be non-existent so that the depletion layer had a well defined edge and the surface-stored charge had zero thickness. In the former case diffusion causes a grading of the majority-carrier concentration into the depletion region so that, at least in principle, the non-zero concentration could pose a recombination 'threat' to the stored minority carriers. The density of majority carriers may be calculated by noting that zero current must flow perpendicular to a clock electrode when the steady state exists.

$$\text{i.e.}\quad pe\mu_p E - eD_p \frac{dp}{dx} = 0 \qquad 2.95$$

p is the majority hole concentration, μ_p is the hole mobility and D_p is the hole diffusion coefficient. Equation 2.95 may be integrated to give

$$\begin{aligned} \ln p &= (\mu_p/D_p)\int E\, dx + \text{constant} \\ &= -(\mu_p/D_p)V + \text{constant} \qquad 2.96 \end{aligned}$$

where V is the electrostatic potential. If $V = 0$ where $p = p_0$ (the undepleted equilibrium concentration) equation 2.96 becomes

$$p = p_0 \exp[-(\mu_p/D_p)V] = p_0 \exp(-eV/kT) \qquad 2.97$$

The Einstein relationship, $D_p/\mu_p = kT/e$, has been used in equation 2.97, which gives the spatial variation of p through the depletion layer if the spatial variation of V is known. At room temperature $kT/e \simeq 25 \times 10^{-3}$ V, so, for example, a surface voltage of only 1 V relative to the substrate will make $p/p_0 \simeq 4 \times 10^{-18}$ which is negligible.

The thickness of the minority carrier layer at an insulator–semiconductor interface may be determined in a similar way from the dynamic equilibrium between diffusion and conduction currents. As was implied in equation 2.26 in considerations of saturation in a surface channel CCD the stored minority carrier density is much greater than the density of the dopant atoms in the semiconductor. Accordingly, E varies significantly through the space charge layer but the approximation $N_A \simeq 0$ in the interface region is allowable so that the zero current requirement and Poisson's equation become respectively

$$ne\mu_n E + eD_n \frac{dn}{dx} = 0 \qquad 2.98$$

$$\text{and} \quad \epsilon_s \epsilon_0 \frac{dE}{dx} = -ne \qquad 2.99$$

Eliminating n from equations 2.98 and 2.99,

$$-\frac{d(E^2)}{dx} = \frac{2D_n}{\mu_n} \frac{d^2E}{dx^2} \qquad 2.100$$

Integrating equation 2.100 with the boundary condition that $E = 0$ when $dE/dx = 0$ gives

$$-E^2 = \frac{2D_n}{\mu_n} \frac{dE}{dx} \qquad 2.101$$

Further integration with $E = -E_2$ at the semiconductor–insulator interface where $x = 0$ gives

$$\frac{E}{E_2} = \frac{-1}{1 - (\mu_n E_2/2D_n)x} \qquad 2.102$$

Equation 2.102 is only applicable in the semiconductor where x is negative. If the potential zero is taken at the interface, further integration of equation 2.102 shows that the voltage, V_{-u}, at a position $x = -u$ is given by

$$V_{-u} = -\int_0^{-u} E \, dx$$

$$= -\frac{2D_n}{\mu_n} \ln\left(1 + \frac{\mu_n E_2}{2D_n} u\right) \qquad 2.103$$

The electron density at any point can be obtained by substituting equation 2.102 in equation 2.99

$$\text{i.e.} \quad n = \frac{n_0}{[1 - (\mu_n E_2/2D_n)x]^2} = \frac{n_0}{[1 + (\mu_n E_2/2D_n)u]^2} \qquad 2.104$$

At $x = 0$

$$n = n_0 = \frac{\mu_n \epsilon_s \epsilon_0}{2D_n e} E_2{}^2 \qquad 2.105$$

Equations 2.104 and 2.102 indicate that $2D_n/\mu_n E_2$ is a characteristic distance, λ_n, for the thickness of the minority charge layer. Equation 2.104 shows that the electron density at a distance λ_n is only a quarter of its value at the interface. Equation 2.99 shows that the total charge per unit area contained in a distance u is

$$\int_{-u}^{0} n \, dx = \frac{\epsilon_s \epsilon_0}{e} (E_{-u} + E_2) \qquad 2.106$$

Therefore half the minority charge is contained in the distance λ_n.

To estimate the value of λ_n under saturation conditions we assume that all the bias voltage exists across the insulator. If $\epsilon_s/\epsilon_I = 4, E_2 = \frac{1}{4} V_B/d$, where d is the insulator thickness. Using the Einstein relation, $D_n/\mu_n = kT/e$,

$$\lambda_n = \lambda_{n,s} = \frac{2D_n}{\mu_n E_2} = \frac{8D_n d}{\mu_n V_B} = 8d\frac{(kT/e)}{V_B} \qquad 2.107$$

For $kT/e = 25 \times 10^{-3}$ V at room temperature and $V_B = 10$ V,

$$\lambda_{n,s} = 0.02\,d \qquad 2.108$$

The magnitude of $\lambda_{n,s}$ and the ratio of the saturation stored charge to the depleted charge discussed earlier imply a minority carrier density $\sim 10^{24}$ m^{-3} in a potential increment across the stored charge of ~0.1 V. Under these conditions the availability of energy levels in the conduction band must be checked. The density of available states in the lowest 0.1 V of the conduction band is $\sim 2.5 \times 10^{26}$ m^{-3} so quantum limitations are not expected to be significant.

At the extreme of small stored charge most of the bias voltage exists across the depletion layer of thickness l.

i.e. $\quad V_B \simeq E_2 l/2 \qquad 2.109$

The characteristic thickness, $\lambda_{n,o}$, of the minority charge under these conditions is given by

$$\lambda_{n,o} = \frac{2D_n}{\mu_n E_2} = \frac{D_n l}{\mu_n V_B} = l\frac{(kT/e)}{V_B} \qquad 2.110$$

For $V_B = 10$ V and $kT/e = 25 \times 10^{-3}$ V,

$$\lambda_{n,o} = 0.0025\,l \qquad 2.111$$

For $V_B = 10$ V, l is usually less than 10 μm (see Fig. 2.3) so for $d = 0.1$ μm,

$$\lambda_{n,o} \lesseqq 0.25d \qquad 2.112$$

Comparison of $\lambda_{n,o}$ and $\lambda_{n,s}$ shows that the stored charge occupies a smaller thickness when the amount of charge is large. This occurs owing to the increase of E_2 caused by the presence of the space charge.

Interest is shown in λ_n because its variation implies that the effective separation of positive and negative charge across the insulator depends on the quantity of stored charge. At first sight, this would make C_{ox} non-linear in equation 2.31 which would have non-linearity implications for the charge-sensing and charge-input techniques considered later. However, the following treatment shows that the effect is to give a voltage offset to the voltage across the insulator rather than to make its capacity non-linear. If the charge stored per unit area in the space $-u$ to 0 is ρ_{-u} and the total charge stored per unit area is ρ_n, equations 2.102 and 2.106 show that

$$\frac{\rho_{-u}}{\rho_n} = 1 + \frac{E_{-u}}{E_2} = 1 - \frac{1}{[1 + (\mu E_2/2D_n)u]}$$

i.e.
$$1 + \frac{\mu E_2}{2D_n}u = \frac{1}{[1 - (\rho_{-u}/\rho_n)]} \qquad 2.113$$

Substitution of equation 2.113 in equation 2.103 and using the Einstein relation, $D/\mu = kT/e$ shows that

$$V_{-u} = -\frac{2kT}{e} \ln\left[\frac{1}{1-(\rho_{-u}/\rho_{n})}\right] \tag{2.114}$$

Equation 2.114 shows that any particular ratio for ρ_{-u}/ρ_n causes V_{-u} to be constant irrespective of the value of ρ_n. Therefore there is a voltage addition, equal to $-V_{-u}$, to the voltage across the oxide when considering the storage of charge. From equation 2.114, twice the thermal voltage, kT/e, contains 63% of the stored charge and ten times the thermal voltage contains 99.3% of the stored charge.

2.14 Electrical charge input to the CCD

So far, details of charge manipulation within the CCD have been discussed without reference to the methods of charge injection or extraction. The input-injection process usually involves analogue control of the charge admitted to the CCD from a large source of electrons by manipulation of the intervening surface- or channel-potential in a manner which has similarities to the source and gate function in a field-effect transistor. The structure and the corresponding energy-band diagrams are illustrated in Figs. 2.29 and 2.30 respectively. The 'infinite' source

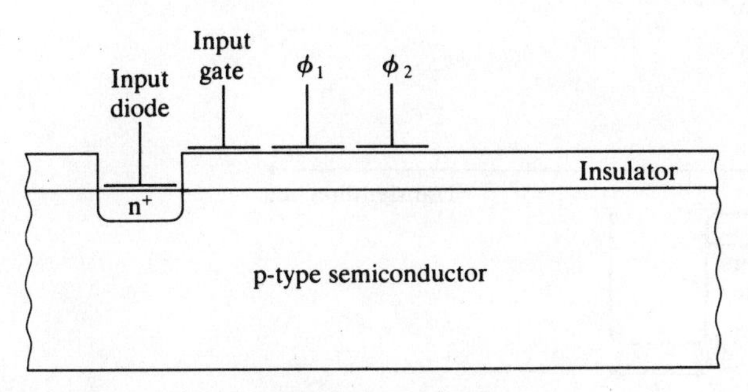

Fig. 2.29 An input structure for a CCD.

of electrons is contained in the conduction band of the n^+ region of the input diode at a potential which is controlled by the magnitude of the reverse bias. The input gate has the same structure as the clock electrodes and is between the input diode and first clock electrode. Electrons will flow from the input diode to the CCD if their potential in the input diode is higher (lower bias voltage magnitude) than the surface potential under the input gate which, in turn, is higher than that of the first clock electrode as shown in Fig. 2.31(a). Raising the surface potential of the input gate (by reducing the magnitude of its positive bias voltage) as in Fig. 2.31(b) will stop the electron flow as will a sufficient reduction of the electron potential energy in the input diode. Similarly electron flow will not

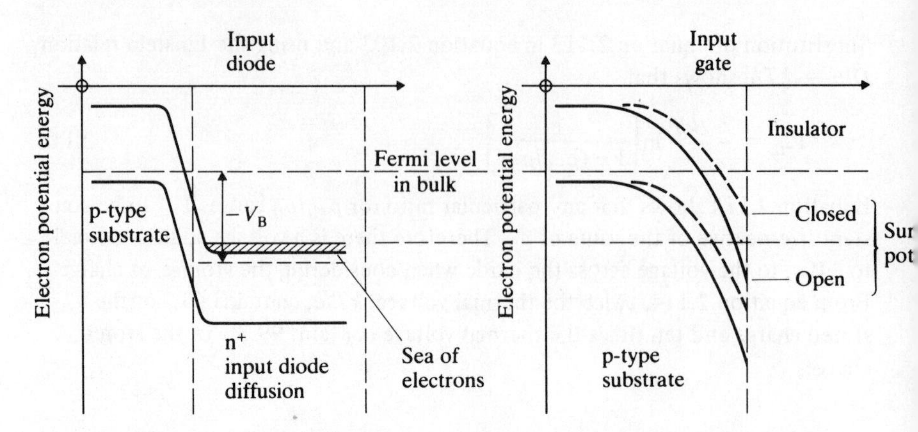

Fig. 2.30 Potential diagrams for the input diode and input gate shown in Fig. 2.29.

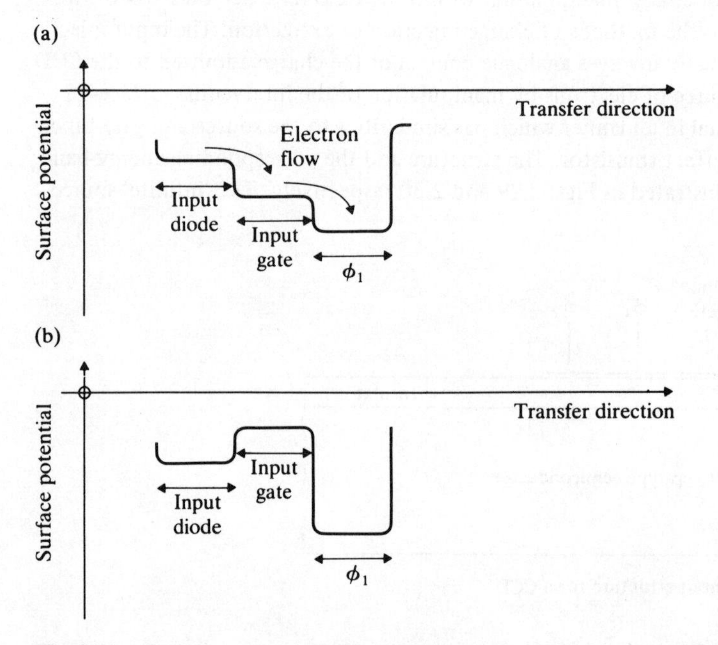

Fig. 2.31 Surface potentials in the input circuit of a CCD: (a) charge injection; (b) charge isolation.

occur if the surface potential under ϕ_1 is higher than the electron potential in the input diode, irrespective of the gate potential. Further details of linearity and timing control of this input circuit will be given in Chapter 5.

2.15 Optical signal injection

If an electron–hole pair is created in the depletion region of an MIS charge store, the minority carrier will fall into the surface-potential minimum and the majority carrier will move into the substrate. Accordingly, photons with an energy greater than the energy-band gap of the semiconductor can be used to create a stored charge which is proportional to the optical intensity. The photons may enter the semiconductor around the edges of the appropriate clock electrode or from the substrate side of the CCD. The preservation of the spatial order of the stored charge in clocking the CCD is naturally compatible with the scan of a television monitor so that this optical effect allows the CCD to act as a television camera or similar imaging device. Greater detail will be given in Chapter 7.

2.16 Electrical charge output from the CCD

The output circuit at the end of the CCD is often similar to the input circuit as illustrated in Fig. 2.32. The output diode is reverse biased so that any minority

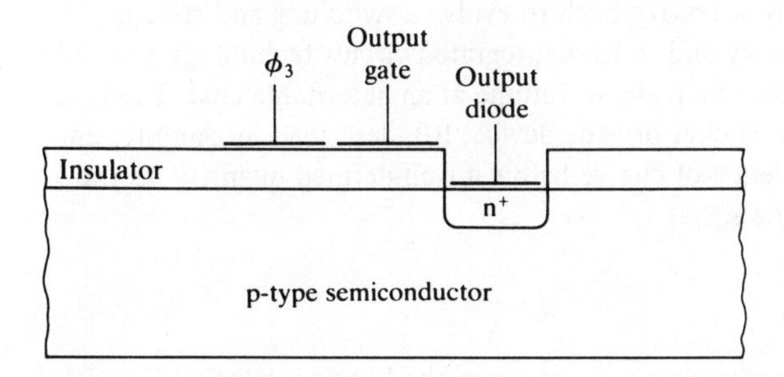

Fig. 2.32 **A destructive output structure.**

electrons entering its depletion region from the CCD flow to the external circuit as an enhancement of the reverse leakage current. The output gate is used to control the surface potential between the ϕ_3 electrode and the output diode so that the signal charge is delivered correctly to the external circuit. It also provides some electrical isolation of the clock pulses on ϕ_3 from the output diode.

In the previous output technique, charge would be completely lost from the CCD during the output process. Many applications require simultaneous sensing of the analogue charge magnitude at a multitude of points along the CCD so that a non-destructive sensing technique is necessary. All the techniques make use of the relationship between charge and voltage at an electrode. The charge on either side of the insulator must be equal in magnitude but of opposite sign. If the electrode is maintained at constant voltage this fact may be used to sense the flow of minority charge into a potential well in the semiconductor region, because there must be a corresponding change in the charge present on the electrode. Alternatively, if the electrode is isolated, minority charge will cause a corresponding reduction in the depletion layer width so that the associated change of electrode voltage may be sensed with a high impedance circuit. Further details of these techniques and a consideration of their linearity will be given in Chapter 5.

3

Elements of Bucket-Brigade Devices

3.1 Introduction

The sampled-data analogue delay line was first made using separately identifiable switches and charge stores. Many realizations are possible using voltage or current instead of charge to carry the signal. In order for the many repeated elements to be incorporated into a practical device, rather than a laboratory curiosity, it was necessary both to evolve a switching and storage element of extreme simplicity and to have integrated-circuit technology available to produce the entire device in a realistic volume at an acceptable cost. The first satisfactory device was the bucket-brigade device, BB, described by Sangster and Teer (1969), using a deficiency of charge below a well-defined quantity, as the means to carry the analogue signal.

3.2 Charge-deficit transfer

One of the switching and storage units is shown in the boxed section of Fig. 3.1. The switching sequence is two-phase and inherent directionality is provided by the transistor switches T_1 and T_2. Capacitors, C_1 and C_2, are the charge stores. The two sections, connected respectively to the clocks ϕ_1 and ϕ_2, in each unit are necessary to isolate successive signals. In Fig. 3.2, V_{c_1}, V_{c_2}, V_{b_1} and V_{b_2} are instantaneous voltages at the collectors and bases of the transistors and the effect is shown of three successive and different signals, V_{s_1}, V_{s_2} and V_{s_3}, on the collector voltage. Even though bipolar transistors are shown in Fig. 3.1 their functions could be carried out with field-effect transistors. Differences in the

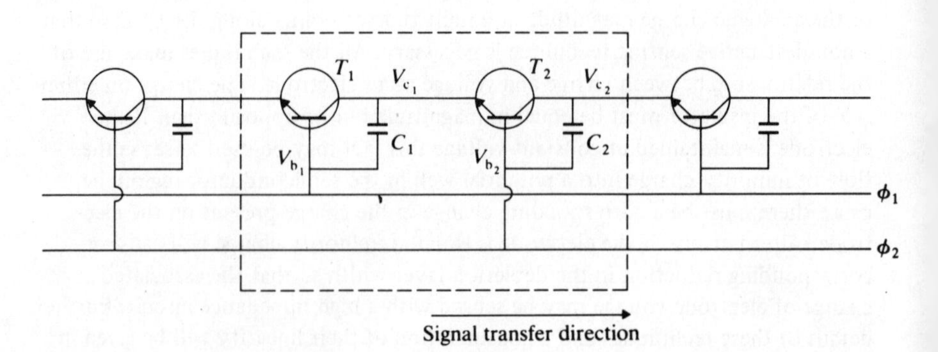

Fig. 3.1 A cell of a bipolar BB.

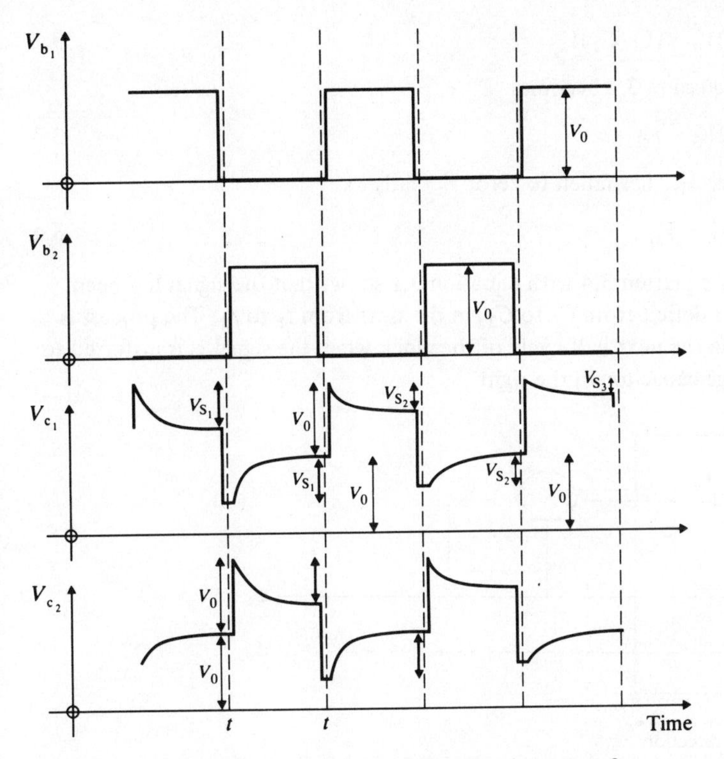

Fig. 3.2 Base and collector waveforms during the charge-transfer process.

details of performance will be described later. The non-overlapping clock waveforms are necessary to isolate adjacent charge stores, in contrast to the CCD where overlapping clock waveforms were necessary.

If we enter the cyclic order of events at time t_1, C_2 has previously been charged up to a voltage equal to the clock voltage amplitude, V_0. A voltage, $V_0 - V_{S_1}$, is stored on capacitor, C_1.

i.e. $V_{C_1} = V_0 - V_{S_1}$ 3.1

In this initial description of the bucket brigade it will be assumed that we use idealized transistors which conduct for positive base-emitter voltage and are open circuit for reverse base-emitter voltage.

Just after t_1, the clock voltage V_0 is applied to the base of T_2 so that V_{C_2} instantaneously rises to $2V_0$. T_2 will now conduct owing to its forward-biased emitter-base junction and positive charge will flow from C_2 to C_1. When $V_{C_1} = V_0$ the emitter-base junction of T_2 becomes zero biased so it ceases to conduct. Throughout this time, the emitter-base junction of T_1 was reverse biased owing to the voltage on the final capacitor of the previous stage, so T_1 provides isolation between simultaneous charge transfers occurring in adjacent cells. This equilibrium state will have been reached when

$$V_{c_2} = 2V_0 - (C_1/C_2)V_{s_1} \quad 3.2$$

If $C_1 = C_2$, equation 3.2 becomes

$$V_{c_2} = 2V_0 - V_{s_1} \quad 3.3$$

at time t_2, after V_{b_2} has fallen to zero, V_{c_2} falls to

$$V_{c_2} = V_0 - V_{s_1} \quad 3.4$$

Comparison of equation 3.4 with equation 3.1 shows that the signal has been transferred as a deficit from C_1 to C_2, in the time from t_1 to t_2. The process is repeated during the next half-cycle of the clock when the signal is transferred to the next storage capacitor to the right.

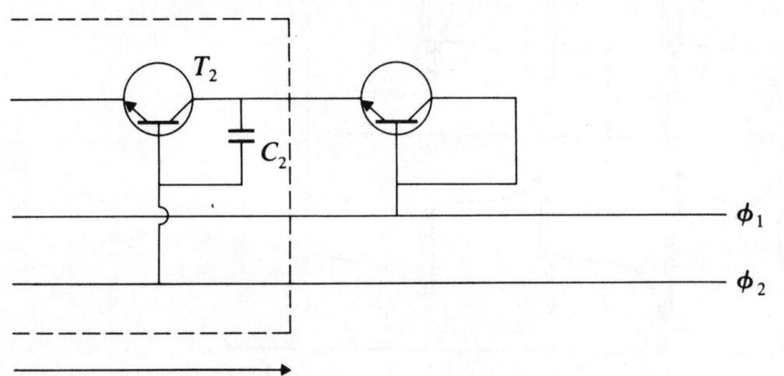

Fig. 3.3 The recharging circuit at the end of the BB.

If any of the V_c ever become less positive than V_b the transfer of positive charge from collector to emitter will not occur. It is necessary to recharge the final capacitor to V_0 after each charge transfer so that there is always sufficient charge in a capacitor to supply the charge deficit in the preceding capacitor. A suitable final stage of the BB is shown in Fig. 3.3.

When the above switching sequence is used it is necessary to arrange that zero signal corresponds to a voltage V_0 stored on the capacitors. This organization is referred to as *charge-deficit transfer*, and is used to avoid charge-transfer difficulties and non-linearity at small signal levels that would be caused by the near-zero value of the collector–emitter voltage during some stages of the transfer process if zero capacitor voltage corresponded to zero signal.

Even though the monolithic integrated-circuit techniques can provide a good match between all the capacitors any residual differences will not affect the amount of charge transfer except at signal levels approaching saturation – when the transistor operating conditions may be violated. For example, if a charge, q, is removed from C_2 to raise V_{c_1} up to V_0 in the above description, the same charge, q, will be required to restore the voltage on C_2 to V_0 one half-cycle later. The only capacitor values that are important in this respect are those at the input and output where the charge is generated from, or converted into, a voltage.

3.3 The sample and hold input circuit

At the input to the BB the signal must be sequentially entered as a charge deficit at a time coincident with the appropriate part of the clock cycle. A typical circuit to carry out this sample and hold function is illustrated in Fig. 3.4. T_1 and C_2 are the first switch and store and are connected to the ϕ_1

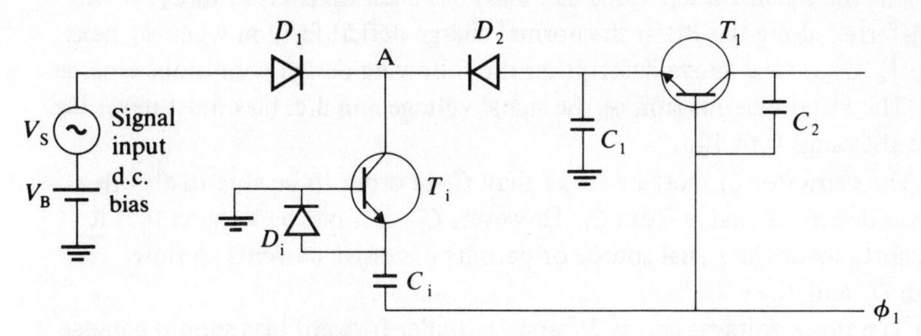

Fig. 3.4 An input sample and hold circuit.

clock phase. When ϕ_1 is on, the voltage across C_1 is returned to the clock voltage amplitude, V_0, by the charge-deficit transfer to C_2. The junction of the emitter of T_i and C_i is clamped to a voltage slightly greater than zero by the forward-biased diode, D_i, so that the emitter–base junction of T_i is reverse biased, Point A is isolated and the oppositely connected diodes D_1 and D_2 isolate C_1 from the low-impedance signal source. Waveforms are illustrated in Fig. 3.5.

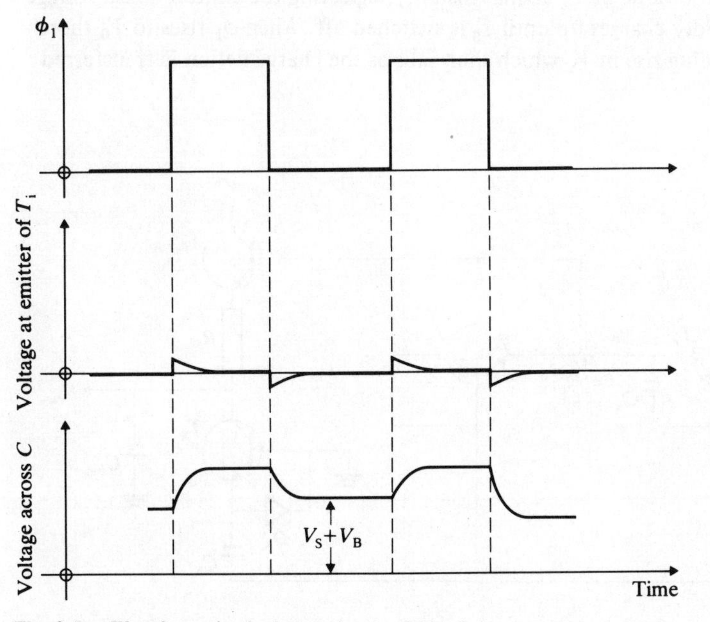

Fig. 3.5 Waveforms in the input circuit of Fig. 3.4.

When ϕ_1 falls to zero T_1 and D_i become non-conducting and T_i is turned on. Charge flows out of C_1, through D_2 and T_i, into C_i until the voltage across C_1 is equal to the sum of the input signal and its d.c. bias. Point A is then clamped at this voltage by the forward bias of D_1 and the low impedance of the signal source. Current will continue to flow through D_1 and T_i until C_i is charged sufficiently to remove the forward bias of the emitter–base junction of T_i. By this means the signal voltage (plus d.c. bias) has been transferred to C_1. It will be transferred along the BB in the normal charge-deficit fashion when ϕ_1 next rises to V_0. From the above description the following design conditions emerge.

(i) The instantaneous sum of the signal voltage and d.c. bias must never lie outside the range 0 to V_0.

(ii) The capacitor C_i must be larger than C_1 in order to be able to absorb a complete deficit of charge from C_1. However, C_i must not be so large that it significantly loads the signal source or permits excessive currents to flow through D_1 and T_i.

(iii) The finite voltages across D_1 and D_2 under forward bias should balance out.

3.4 Signal tapping, the output circuit and signal reconstruction

Circuitry typically used to sense the signal non-destructively is illustrated in Fig. 3.6. It may be used at any stage of the BB or at the output as illustrated. The time variation of V_X in Fig. 3.7 has the same form as any equivalent charge store in the BB. T_S, D_S and C_S perform similar operations to T_i, D_i and C_i at the input. When ϕ_2 falls to zero, T_S is turned on and discharges C_{out} until V_{out} is equal to V_0, (the value of V_X at that instant, neglecting the emitter–base voltage of T_e). C_S rapidly charges up until T_S is switched off. When ϕ_1 rises to V_0 there is a corresponding rise in V_X which then falls as the charge deficit is transferred

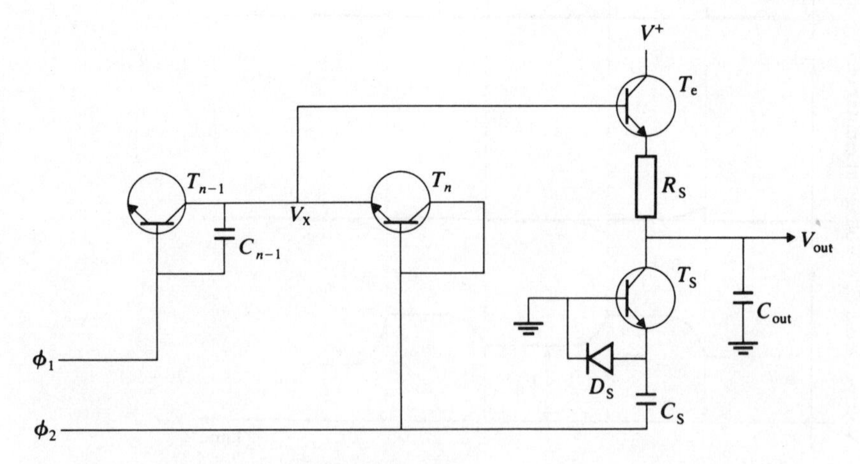

Fig. 3.6 An output (or signal tapping) circuit.

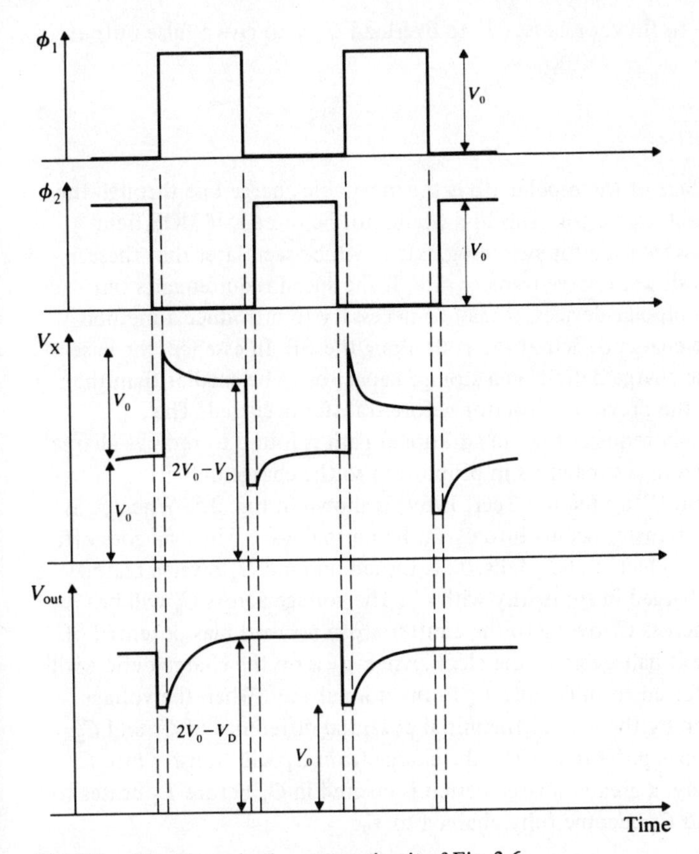

Fig. 3.7 Waveforms in the output circuit of Fig. 3.6.

from C_{n-2}. The resistor R_s has been schematically included in Fig. 3.6 to identify the relatively slow charging of C_{out} compared with the fall of V_X, so that V_{out} asymptotically approaches the desired signal voltage (plus a d.c. offset). This voltage is held when ϕ_1 falls to zero and when ϕ_2 rises to V_0 because the emitter–base junction of T_e is reverse biased until the next time that T_s is switched on by ϕ_2 falling to zero. The voltage across C_s is reset by D_s when ϕ_2 rises to V_0. In practice, R_s is an undesirable component for an integrated-circuit process and is not needed because T_{n-1} is operating in common-base mode just after ϕ_1 is applied and T_e is acting in the slower common-emitter mode.

As can be seen from the form of V_{out} in Fig. 3.7, a long clock period compared both with the charging time constant of C_{out} through T_e, and with the separation of ϕ_1 and ϕ_2, results in a signal which is largely reconstructed into its analogue form. For correct operation V^+ must be greater than $2V_0$ and C_s must be greater than C_{out} so that it can completely discharge C_{out} at the appropriate time in one clock cycle. C_s must not be so large that it causes excessive current to flow, or causes T_e to overload C_{n-1} to give a false output signal.

3.5 Charge amplifiers

A great disadvantage of the bipolar BB is the inevitable charge loss through the base current of each transistor. This loss would not be present if MOS field-effect transistors were used for switching, but it will be seen later that these transistors have a slower charge-transfer rate. If the speed requirements can only be met with bipolar devices, it may be necessary to introduce a method of amplifying the charge deficit at intervals along the BB. In essence, the base current causes the charge deficit on a storage capacitor to be smaller than the charge deficit on the previous capacitor before transfer occurred. The amplification simply requires that an additional path is found to remove charge from one of the storage capacitors in proportion to the charge deficit.

One such circuit (Sangster and Teer, 1969) is shown in Fig. 3.8. When ϕ_1 is on, charge-deficit transfer occurs into C_1 in the normal way. When ϕ_1 goes off, the voltage at the collector of T_1 falls, T_a is turned on (and D_a reverse biased) so that C_a is discharged in sympathy with C_1. The voltage across C_a will be larger than that across C_1 owing to the emitter–base forward bias potential of T_a. During the next half-cycle of the clock, when ϕ_2 is on the charge deficit will initially be transferred from C_2 into C_1 in the normal way. When the voltage across C_1 has risen by the sum of the initial potential difference of C_1 and C_a and the forward bias potential of D_a, the charge from C_2 will transfer into C_1 and C_a. In this way, a greater charge deficit is created in C_2 before T_2 ceases to conduct as C_1 and C_a become fully charged to V_0.

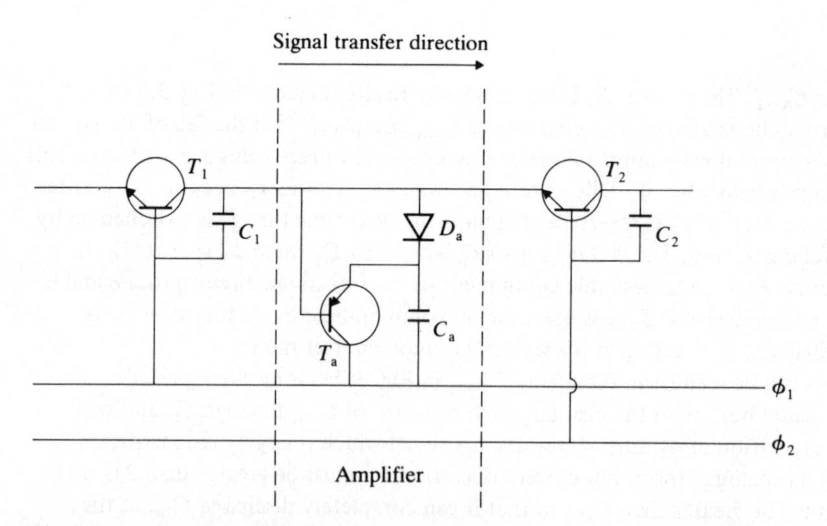

Fig. 3.8 A charge amplifier circuit.

To calculate the charge gain, we note that a charge deficit, Q_S, causes a voltage change, V_S, across C_1 given by

$$V_S = Q_S/C_1 \tag{3.5}$$

By the nature of the charge-transfer process, V_S must be reduced to zero by transfer of a charge, Q_S', from C into $C_1 + C_a$.

$$\therefore \quad Q_S' = (C_1 + C_a)V_S \tag{3.6}$$

From equations 3.5 and 3.6 the charge amplification factor, A_Q, is

$$A_Q = Q_S'/Q_S = (C_1 + C_a)/C_1 \tag{3.7}$$

The required magnitude of C_a relative to C_1 will be determined by the charge loss. If the common-emitter current gain of each transistor is β, the charge loss after n stages of transfer is A_n, which is given by

$$A_n = [1 - (1/\beta)]^n \tag{3.8}$$

The condition for no net charge loss is

$$A_n A_Q = 1 \tag{3.9}$$

If it is assumed that a charge amplifier is introduced before the signal is significantly reduced (this will be necessary if signal/noise considerations are important) equation 3.9 combined with equations 3.7 and 3.8 approximates to

$$C_a/C_1 = n/\beta \tag{3.10}$$

A few charge amplifiers will be required in every β stages of the BB to avoid A_n becoming too small. In practical terms equation 3.8 implies an attenuation of 8.68 dB after every β transfers or approximately 3 dB after every $\beta/3$ transfers.

A disadvantage of this amplifier is the necessity to transfer a minimum d.c. charge deficit in order for C_a to ever enter the charge-transfer process. Otherwise serious distortion would occur at small signal levels. Unfortunately this effect reduces the dynamic signal range of the device. The offset can be halved by making T_2 as a transistor with two emitters, one directly connected to C_a.

3.6 *p*-phase clocking and an increase in packing density

Only half of the BB contains a signal charge deficit for the two-phase device. More efficient use may be made of the semiconductor chip area if more complicated clock systems are used. The techniques are directly analogous to the electrode–bit schemes discussed for CCD. For example, the three-phase scheme illustrated in Fig. 3.9 contains two signal bits in each three transistor–capacitor cell. Initially, the voltage across C_3 is V_0, the voltage across C_2 is $V_0 - V_{S_1}$ and the voltage across C_1 is $V_0 - V_{S_2}$. Application of ϕ_3 causes a transfer of the charge deficit from C_2 to C_3 so that the voltage across C_3 is $V_0 - V_{S_1}$ and the voltage across C_2 is V_0. ϕ_3 is then reduced to zero and is followed by application of ϕ_2 so that the charge deficit on C_1 is transferred to C_2. Repetition of this process with ϕ_1 completes the three-phase process, with the signal initially on C_2 having been transferred to C_4, the signal initially on C_1 having been transferred to C_2 and the signal initially in the cell which is two elements before C_1 having been transferred to C_1. Generalization of these ideas to a p-phase device shows that the fraction of transistor–capacitor cells containing a signal is $[(p-1)/p]$.

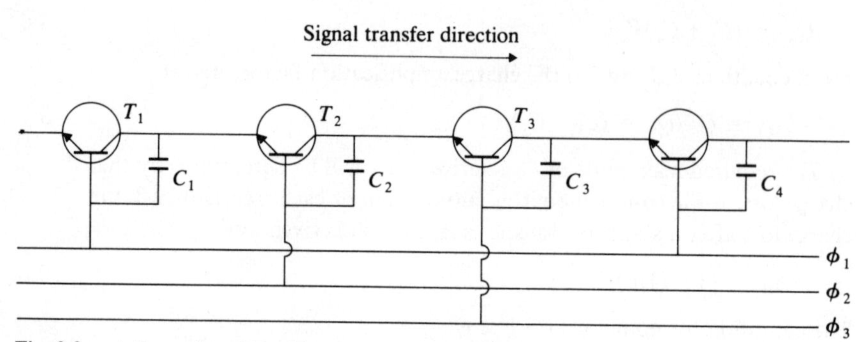

Fig. 3.9 A three-phase BB with greater packing density.

In addition to the disadvantage of complexity in the clock driving circuitry for p-phase devices the time available for charge-transfer decreases with increasing p as shown by the following argument. If N bits of information must be stored for a time T_d there must be M transistor–capacitor cells where

$$N = [(p-1)/p]M \qquad 3.11$$

$(p-1)$ clock repetition periods, T_c, are required to transfer the charge along p cells of the BB.

$$\text{i.e.} \quad T_d = [(p-1)/p]M\,T_c \qquad 3.12$$

In each T_c there are p transfer times, T_T.

$$\text{i.e.} \quad p\,T_T = T_c \qquad 3.13$$

Combining equations 3.11, 3.12 and 3.13,

$$T_T = \frac{T_d}{N}\frac{1}{p} \qquad 3.14$$

Therefore the time available for transfer is inversely proportional to p. Later considerations of the charge-transfer time will imply that the decrease of T_T with increasing p may limit the accuracy of the charge-deficit transfer.

3.7 The charge-transfer rate

3.7.1 *Transconductance effects*

The rate of charge transfer is dominantly controlled by the gate-source voltage of a field-effect transistor or by the base–emitter junction of a bipolar transistor in the circuit of Fig. 3.10. In both cases the relationship between the controlling voltage, $V_{1,2}$, and the current, i, is non-linear so that i decreases more rapidly than $V_{1,2}$ causing a charge-transfer rate that decreases with time.

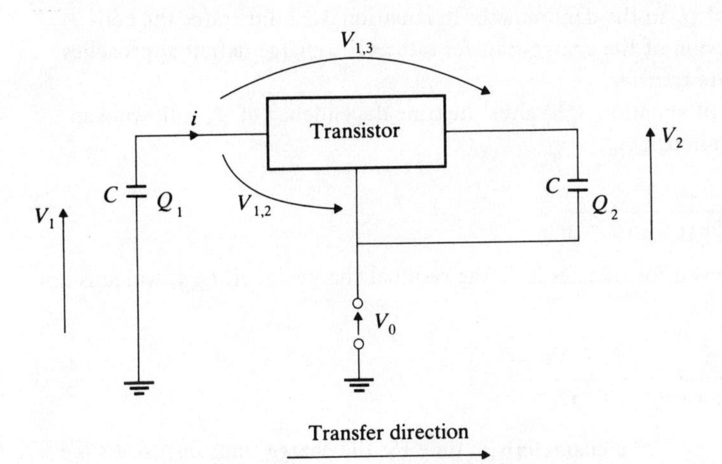

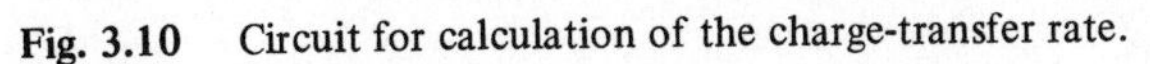

Fig. 3.10 Circuit for calculation of the charge-transfer rate.

(i) The IGFET BB Ideally, the insulated-gate field-effect transistor (IGFET) has a square-law relationship between i and $V_{1,2}$ (Sze, 1969):

$$i = \gamma(V_{1,2} - V_T)^2 \tag{3.15}$$

V_T is the threshold voltage for conduction and γ is a constant.

The signal charge deficit, Q_S, on the input capacitor is given by

$$Q_S = C(V_0 - V_T) - Q_1 \tag{3.16}$$

i.e. $$Q_S = C(V_0 - V_T - V_1) \tag{3.17}$$

where V_0 is the clock-voltage amplitude and Q_1 is the charge on C_1. i and $V_{1,2}$ are given by

$$i = \mathrm{d}Q_1/\mathrm{d}t \tag{3.18}$$

$$V_{1,2} = V_0 - V_1 \tag{3.19}$$

Substitution of equations 3.18 and 3.19 into equation 3.15 and elimination of Q_1 and V_1 with equations 3.16 and 3.17 gives

$$\frac{\mathrm{d}Q_S}{\mathrm{d}t} = -\gamma\left(\frac{Q_S}{C}\right)^2 \tag{3.20}$$

This equation has the same form as the self-induced field process in a CCD as described in equation 2.75. Rewriting equation 3.20 in the form

$$\frac{1}{Q_S}\frac{\mathrm{d}Q_S}{\mathrm{d}t} = -\frac{\gamma Q_S}{C^2}, \tag{3.21}$$

shows that the instantaneous charge-transfer time constant, τ_γ, in an expression of the form Q_s = constant x $\exp(-t/\tau_\gamma)$, is given by

$$\tau_\gamma = C^2/\gamma Q_s \tag{3.22}$$

The presence of Q_s in the denominator in equation 3.22 illustrates the considerable reduction of the charge-transfer rate as the charge deficit approaches zero at complete transfer.

Integration of equation 3.20 gives the time dependence of Q_s following an initial charge deficit, $Q_{s,o}$,

$$\frac{Q_s}{Q_{s,o}} = \frac{1}{1 + (Q_{s,o}\gamma/C^2)t} \tag{3.23}$$

If the time allowed for transfer is T_c the residual charge deficit $Q_{s,r}$, which is not transferred is

$$\frac{Q_{s,r}}{Q_{s,o}} = \frac{1}{1 + (\gamma Q_{s,o}/C^2)T_c} \tag{3.24}$$

From equation 3.23 a characteristic time for the charge-transfer process is τ_γ, given by

$$T_\gamma = C^2/\gamma Q_{s,o} \tag{3.25}$$

The maximum value of $Q_{s,o}$ is $Q_{s,\max}$ which is approximately equal to CV_0 where V_0 is the clock voltage amplitude. For typical values of $C = 10^{-12}$ F, $\gamma = 5 \times 10^{-5}$ A V^{-2} and $V_0 = 10$ V

$$T_\gamma = 2 \times 10^{-9}\,\text{s} \quad \text{for } Q_{s,o} = Q_{s,\max} \tag{3.26}$$

and $$T_\gamma = 2 \times 10^{-8}\,\text{s} \quad \text{for } Q_{s,o} = 0.1\,Q_{s,\max} \tag{3.27}$$

T_c must be much longer than T_γ to allow a substantially complete charge-deficit transfer. For the values given in equations 3.26 and 3.27 this condition will be met for clock rates below a few MHz. To first order equation 3.26 then becomes

$$Q_{s,r} \simeq \frac{C^2}{\gamma T_c}\left(1 - \frac{C^2}{Q_{s,o}\gamma T_c}\right) \tag{3.28}$$

For $T_c > 10^{-6}$ s and $Q_{s,o} > 0.1\,Q_{s,\max}$, equations 3.26 and 3.27 show that the term $\dfrac{C^2}{Q_{s,o}\gamma T_c}$ in equation 3.28 is less than 0.02. Equation 3.28 shows that $Q_{s,r}$ is not linearly dependent on $Q_{s,o}$ and it is unfortunate that this non-linearity becomes more pronounced for smaller $Q_{s,o}$.

If $Q_{s,r}$ is not negligible for all the desired dynamic range, linear operation requires it to be constant for all values of $Q_{s,o}$ causing a simple d.c. offset. Alternatively, the ratio $Q_{s,r}/Q_{s,o}$ must be constant. The former option is the only one open. To reduce the signal dependence of $Q_{s,r}$ a charge offset (fat zero) must be added to all signals at the input of the BB. It has the disadvantage of correspondingly reducing the dynamic range of signals that can be transferred through the BB. Equation 3.28 shows that an offset equal to 0.1 $Q_{s,\max}$ will only allow a 2% variation of $Q_{s,r}$ for all signal conditions at $T_c = 10^{-6}$ s. Also $Q_{s,r}$ is only 2% of $Q_{s,\max}$. This suggests that the effective charge-transfer inefficiency in the present case is 4×10^{-4} per transfer.

All the above numerical estimates were made for a silicon IGFET with an n-type channel. Slower transfer rates will occur in a silicon IGFET with a p-type channel owing to the direct dependence of γ on the mobility of the hole-charge carriers which is approximately 40% of that for electrons in an n-channel IGFET (Sze, 1969).

It should be pointed out also that the residual charge fraction will be smallest at the highest possible clock voltage owing to the direct dependence of the saturation charge deficit on the clock voltage. Further details of these considerations and design details are given by Berglund and Strain (1972) and by Berglund and Boll (1972).

(ii) The bipolar BB A similar behaviour occurs in the bipolar BB where the relationship between i and $V_{1,2}$ is

$$i = i_0 [\exp(eV_{1,2}/kT) - 1] \tag{3.29}$$

For the present purposes the charge-deficit loss treated in section 3.5 will be neglected.

Using equations 3.16, 3.17, 3.18 and 3.19 (with $V_T = 0$ for a bipolar transistor) in equation 3.29,

$$-\frac{dQ_s}{dt} = i_0\{\exp[(e/kT)(Q_s/C)] - 1\} \tag{3.30}$$

Rewriting equation 3.30,

$$\frac{-\exp[-(e/kT)(Q_s/C)]\, dQ_s}{\{1 - \exp[-(e/kT)(Q_s/C)]\}} = i_0 dt \tag{3.31}$$

Integrating equation 3.31,

$$Q_s = -\frac{kT}{e} C \ln \left(1 - \exp\left\{\frac{-t}{(kT/e)(C/i_0)}\right\} + \exp\left\{\frac{-[(Q_{s,0}/i_0) + t]}{[(kT/e)(C/i_0)]}\right\}\right) \tag{3.32}$$

The residual charge is independent of the initial charge if

$$Q_{s,0} \gg (kT/e)C \tag{3.33}$$

As in the case of the IGFET BB a constant charge deficit must be added to the signal to avoid signal distortion. However, a standing charge-deficit transfer of ten times $(kT/e)C$ should amply satisfy equation 3.33 and will only cause a voltage offset of approximately 250 mV at room temperature, so that the dynamic range of the bipolar BB will not suffer as great a restriction as was the case for the IGFET BB. This difference in performance essentially arises from the exponential characteristic of the idealized bipolar BB which causes a very rapid initial transfer of charge. A more important mechanism in the bipolar BB is the charge-loss mechanism discussed in section 3.5, and the charge feed-forward described in section 3.8.

3.7.2 Dynamic drain conductance

Equation 3.28 showed that the presence of a standing charge deficit could make the residual charge deficit independent of the initial charge deficit to an accuracy of approximately 2% for the conditions quoted.

Unfortunately a further complication arises from the increase of γ with the drain to source voltage, V_{ds}, which is caused by reduction of the IGFET channel length with increasing V_{ds} as the drain depletion layer broadens underneath the gate and extends towards the source (Sze, 1969). As shown in Fig. 3.11 this

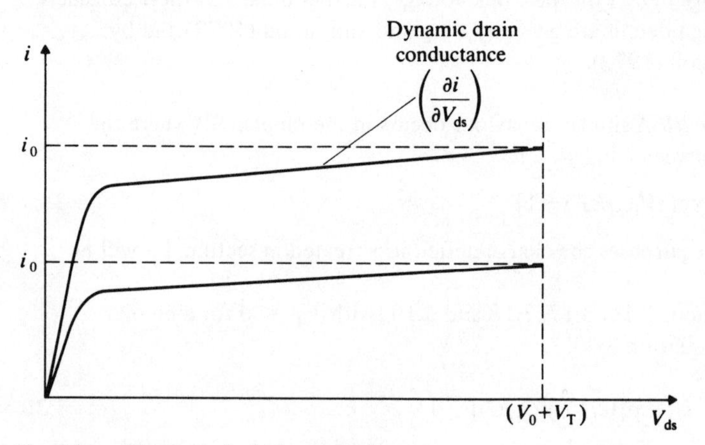

Fig. 3.11 The dynamic drain conductance.

effect causes a finite output conductance which is known as the dynamic drain conductance. In turn γ is made dependent on $Q_{s,0}$ because when transfer is substantially complete V_{ds} is given by

$$V_{ds} = [V_0 - (Q_{s,0}/C)] + V_T \tag{3.34}$$

If an expression of the form of equation 3.15 is used to describe curves of the form shown in Fig. 3.11, above the knee voltage, γ must have the form

$$\gamma = \gamma_0 \left[1 - \frac{1}{i_0}\left(\frac{\partial i}{\partial V_{ds}}\right)(V_0 + V_T - V_{ds})\right] \tag{3.35}$$

γ_0 is the value of γ corresponding to i_0, the drain current for zero charge deficit. Equation 3.35 is useful because $(1/i_0)(\partial i/\partial V_{ds})$ is often approximately constant for any particular curve in Fig. 3.11 (i.e. for a particular value of $V_{1,2}$ in equation 3.15). If it is assumed that the dominant effect is to modify the charge-transfer rate in the slow final stages of transfer, equation 3.35 can be written in the approximate form

$$\gamma \simeq \gamma_0 \left[1 - \frac{1}{i_0}\left(\frac{\partial i}{\partial V_{ds}}\right)\frac{Q_{s,0}}{C}\right] \qquad 3.36$$

Numerical estimates of this effect depend on the detail of the IGFET design but a 10% change in γ over the full working range of $Q_{s,0}$ is not unreasonable. This variation of γ causes a corresponding variation of $Q_{s,r}$ in equation 3.28. Accordingly, the dynamic drain conductance causes a more serious residual charge deficit than that already calculated from the simple charge-transfer rate in the presence of a fat zero. The usual design technique to minimize the dynamic drain conductance is to use a high doping level underneath the IGFET gate so that the dependence of the drain depletion layer width on drain voltage is minimized. In turn this reduces the dependence of the effective IGFET channel length on the drain voltage.

3.7.3 Standing current

As shown by equation 3.21 the rate of charge transfer decreases as Q_s tends to zero. In the above discussion the dependence of the residual charge deficit on the initial charge deficit was reduced by adding a standing level of charge deficit to the signal. This is not really solving the problem which is caused by i, in equation 3.15, tending to zero as $V_{1,2}$ approaches V_T. The final stages of charge transfer would be much more rapid if the process terminated with a finite i instead of zero i. This may be achieved by applying a trapezoidal clock waveform, as illustrated in Fig. 3.12, to the gate of the transistor. Equation 3.20 is then modified to

$$\frac{dQ_s}{dt} = -\gamma\left(\frac{Q_s}{C}\right)^2 + \frac{C\,dV_0}{dt} \qquad 3.37$$

dV_0/dt is a constant.

From equation 3.37 the charge-transfer process terminates at a finite $Q_s = Q_{s,f}$ given by

$$Q_{s,f} = \left(\frac{C^3}{\gamma}\frac{dV_0}{dt}\right)^{\frac{1}{2}} \qquad 3.38$$

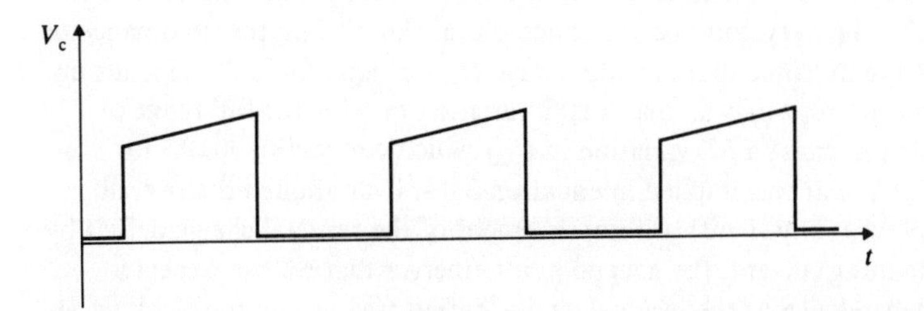

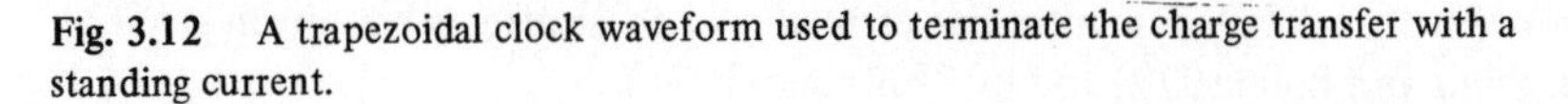
Fig. 3.12 A trapezoidal clock waveform used to terminate the charge transfer with a standing current.

Rewriting equation 3.37, the instantaneous time constant, $\tau_{\gamma s}$, is given by

$$\frac{1}{\tau_{\gamma,s}} = \frac{1}{Q_s}\frac{dQ_s}{dt} = -\frac{\gamma Q_s}{C^2} + \frac{C}{Q_s}\frac{dV_o}{dt} \qquad 3.39$$

In the initial stages of charge transfer the transfer rate is not significantly affected by the standing current. A high charge-transfer rate is maintained until Q_s is very close to $Q_{s,f}$ when the transfer rate cuts off much more rapidly than it does with conventional operation. For example, the time constant $\tau_{\gamma,f}$, when $Q_s \cong 1.6\, Q_{s,f}$ is given by

$$\tau_{\gamma,f} = \frac{C^2}{\gamma Q_{s,f}} = \left[\frac{C}{\gamma}\frac{1}{(dV_o/dt)}\right]^{\frac{1}{2}} \qquad 3.40$$

Using the previous parameters, $C = 10^{-12}$ F, $\gamma = 5 \times 10^{-5}$ A V^{-2} and assuming $dV_o/dt = 5$ V μ^{-1} we have

$$\tau_{\gamma,f} = 6.3 \times 10^{-8} \text{ s} \qquad 3.41$$

The corresponding value of $Q_{s,f}$ from equation 3.38 is

$$Q_{s,f} = 3.16 \times 10^{-13} \text{ F} \qquad 3.42$$

In equation 3.42 $Q_{s,f}$ is 1/30 of the saturation charge for a clock voltage of 10 V. In equation 3.41, $\tau_{\gamma,f}$, is only 30 times longer than the instantaneous time constant for $Q_s = Q_{s,\max}$.

The magnitude of the residual charge, $Q_{s,r}$, after a finite charge-transfer time, T_c, can be obtained by integrating equation 3.37

$$\text{i.e.} \quad \frac{(Q_{s,r} - Q_{s,f})}{(Q_{s,r} + Q_{s,f})} = \frac{(Q_{s,o} - Q_{s,f})}{(Q_{s,o} + Q_{s,f})}\exp\left(-\frac{2\gamma Q_{s,f}T_c}{C^2}\right) \qquad 3.43$$

In equation 3.43, the factor $\frac{(Q_{s,o} - Q_{s,f})}{(Q_{s,o} + Q_{s,f})}$ lies between 0 and 1, dependent on the initial charge deficit. Also, $\frac{C^2}{2\gamma Q_{s,f}} = \frac{\tau_{\gamma,f}}{2} = 3.2 \times 10^{-8}$. For $T_c > 10^{-8}$ s and $Q_{s,o}$ somewhat greater than $Q_{s,f}$ the factor on the right-hand side of equation 3.43 is less than 2.5×10^{-14}.

$$\text{i.e.} \quad Q_{s,r} - Q_{s,f} < 5 \times 10^{-14} \qquad 3.44$$

Inequality 3.44 shows the maximum variation in the residual charge deficit. Unfortunately, this very good performance is again spoiled by the dominance of the effect of the dynamic drain conductance. $Q_{s,f}$ in equation 3.38 depends on the inverse square root of γ so that a 10% variation in γ for the full range of charge deficit will cause a 5% variation in $Q_{s,f}$ which completely masks the accuracy of $Q_{s,r}$ that was implied in equation 3.44. Even though the overall effect is to prevent any significant improvement of the residual charge deficit by the use of standing current, the general point emerges that we can expect a significant dependence of the residual charge-deficit fraction on the clock waveform. Further detailed quantitative estimates of these effects have been given by Berglund and Boll (1972) and by Thornber (1971).

In all the above estimates of the residual charge deficit, the overriding problem has been the dependence of the final transfer rate on the charge deficit already transferred. This has essentially been caused by the separation of the switching and storage functions into separate devices. The problem does not exist in a well-designed CCD, such as that incorporating polysilicon technology described later. The final transfer rate is only determined by the remaining charge without influence from the charge already transferred.

3.8 Tetrode structures

The separate definition of transistors in a BB inevitably results in stray capacitive couplings that are not intended in the initial design. The stray drain to source capacities in an IGFET BB, as shown in Fig. 3.13, cause an effect similar to

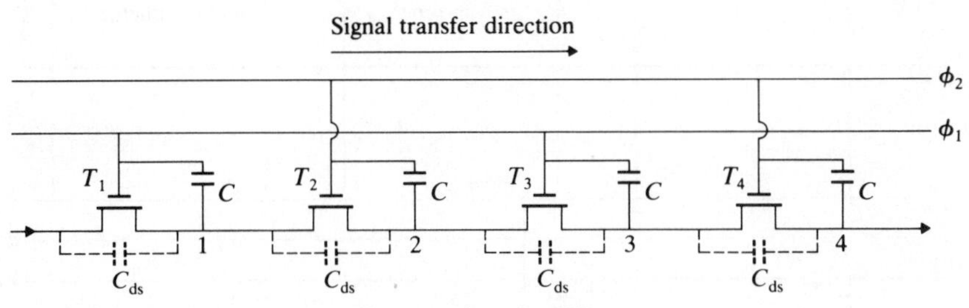

Fig. 3.13 Spurious charge-coupling capacitors in a BB.

charge-transfer inefficiency except that the spurious charge transfer is forward as well as trailing. A similar effect occurs between the collector and emitter of a bipolar bucket brigade. If ϕ_2 in Fig. 3.13 is on and ϕ_1 is off, charge-deficit transfer will be taking place at positions 2 and 4, while the voltage at positions 1 and 3 will be asymptotically defined by transistors T_2 and T_4 respectively. In effect, the storage capacitor receiving the charge deficit is C, in parallel with two spurious C_{ds}s. If the ultimate charge deficit transferred to position 2 is Q_{S_2}, a proportion of it, Q_{ds}, is stored on the spurious capacitive connection between positions 2 and 3.

$$\text{i.e.} \quad Q_{ds} = \frac{C_{ds}}{(C + 2C_{ds})} Q_{S_2} \qquad 3.45$$

There must also be charge deficit equal to Q_{ds} at position 3 on the spurious capacity, and this has to be provided from the storage capacity at position 4, owing to T_4 being in the transfer mode with a defined voltage at position 3. In a typical integrated circuit realization of the IGFET it is difficult to make $C_{ds} < 0.001$ C without making C so large that it seriously increases the charge-transfer time.

One solution to this problem is to improve the isolation of signals between C and T_3 at point 2 in Fig. 3.13. This can be done with another IGFET in the so-called tetrode structure illustrated in Fig. 3.14. The gate of the additional transistor is connected to a d.c. bias and the circuit has similarities to a conventional cascode circuit. When a charge deficit is being transferred to position 2 in Figs. 3.13 and 3.14, the additional transistor in the later case is turned off so that it isolates the spurious signal fed forward in Fig. 3.13. However, the isolation is not complete because the additional transistor also has a

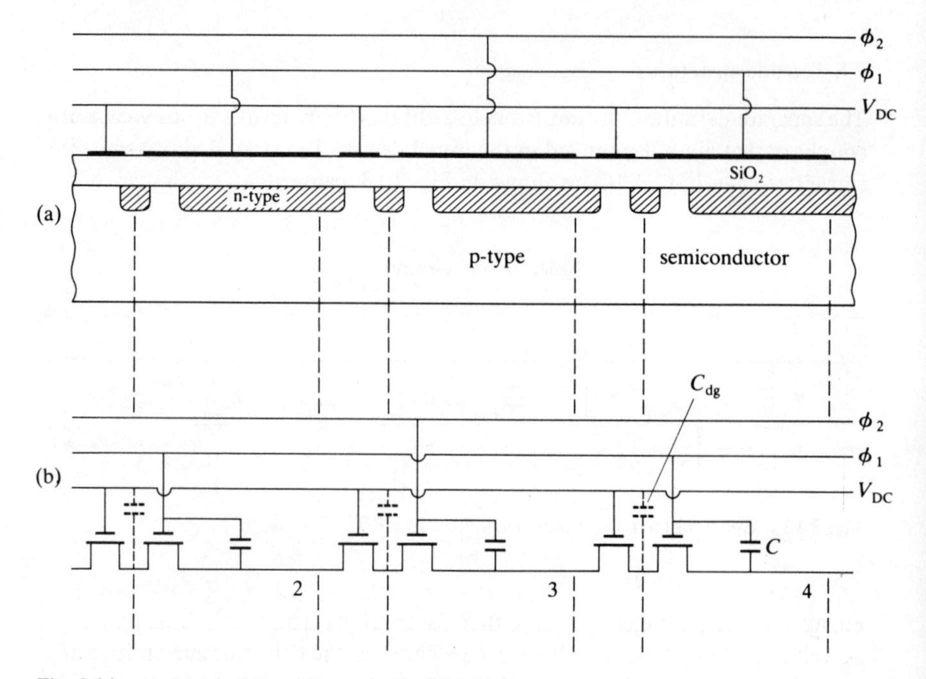

Fig. 3.14 A tetrode BB and its equivalent circuit.

spurious drain to source capacity, and a spurious drain to gate capacity, C_{dg}. The spurious signal in equation 3.45 is then attenuated by a factor C_{ds}/C_{dg} which can be as small as 0.001 as noted above. The implied decrease of the residual charge deficit to $\sim 10^{-6}$ is not completely achieved in practice because each extra transistor and its associated C_{dg} require an extra charge-transfer event which leaves a residual charge deficit. In turn this implies a lower maximum clock frequency to allow more complete charge transfer. Boostra and Sangster (1972) have given more details of this compromise and have reported a tenfold reduction of the maximum clock frequency (to ~50 Hz) for a construction in which a C_{dg}/C of 0.05 gave a twentyfold reduction of the charge smearing when compared with the simple IGFET BB. However, these technological problems have been overcome in devices now commercially available with clock frequencies in excess of 1 MHz.

4

Charge-Transfer Defects

4.1 Incomplete charge transfer

The ideal shifting of a sampled signal is modified by incomplete charge transfer in any practical device. Some origins of this behaviour have been seen in sections 2.12, 3.7 and 3.8, where the finite transfer time restricted the amount of charge transfer, and in section 3.5 where the finite gain of a bipolar transistor caused charge loss. In the former case some of the signal charge was 'smeared' from one cell to the next but in the latter case the signal was permanently lost. Further defects arise in any practical charge-transfer device owing to the inevitable presence of imperfections of semiconductor technology. Initially the sources of these charge-transfer defects will be described. This will be followed by discussions of the effects on charge-transfer device operation and techniques will be considered for their minimization.

4.2 The effect of surface states

4.2.1 Charge smearing

In any real semiconductor there are crystal imperfections and chemical impurities which create extra electron energy levels. Their concentration is particularly high at interfaces such as the $Si-SiO_2$ interface of a surface channel CCD or BB MOS capacitor. The spurious energy levels causing trouble lie in the forbidden energy gap. They are localized in space and any charge carriers falling into them cannot move when required to do so by the charge-transfer process. These spurious levels are usually referred to as surface states (or fast states) and a typical distribution of their concentration in the forbidden energy gap is illustrated in Fig. 4.1. Reviews of their origin, their dependence on crystallographic orientation and technological procedures to minimize their concentration have been given by Sze (1969) and Goetzberger (1974).

During the temporary storage of charge in a potential well or MOS capacitor some of it falls into the surface states. When charge transfer takes place this immobile, or trapped, charge is released at a rate depending on the particular surface states. Some of it will not be released in the available time interval and it becomes part of the following charge packet. The effect will be particularly bad if the analogue signal has a zero or a small sample immediately following a large sample.

In order to evaluate quantitatively the effect of surface states on charge smearing it is necessary to calculate the rates of capture and release of charge carriers. The physical considerations are similar to those for recombination and generation of electrons and holes (Sze, 1969) respectively, and the time constant is

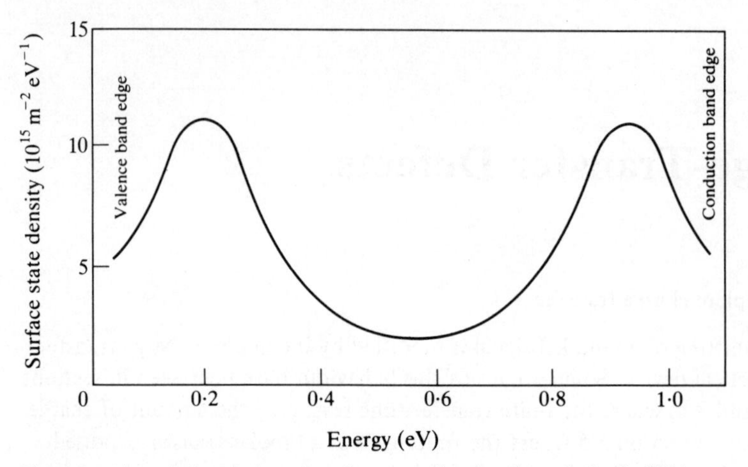

Fig. 4.1 A typical variation of surface state density through the forbidden energy gap of silicon at the Si–SiO_2 interface.

essentially determined by the time required for a carrier to collide with a surface state. The exponential time constant is given by

$$\tau_{\text{fill}} = 1/\sigma v n \tag{4.1}$$

where σ is the capture cross-section area, v is the mean thermal speed and n is the minority electron density (Carnes and Kosonocky, 1972; Lamb *et al*, 1974). Equation 4.1 is usually written in the form

$$\tau_{\text{fill}} = \lambda_{n,s}/\sigma v n_s \tag{4.2}$$

where $\lambda_{n,s}$ is the thickness of the surface charge layer and n_s is the free charge density per unit area. In section 2.13 it was shown that $n \simeq 10^{24}$ m^{-3} under saturation conditions so that with $\sigma = 10^{-19}$ m^{-2} (Lamb *et al*, 1974) and $v = 10^5$ ms^{-1}, the shortest filling time constant is given by

$$\tau_{\text{fill}} \simeq 10^{-10}\ \text{s} \tag{4.3}$$

When the charge-transfer process is taking place, n becomes zero and only the emptying process may occur. The trapped carrier has to be thermally excited to the appropriate free carrier band. The time constant is

$$\tau_{\text{empty}} = \frac{1}{\sigma v N_c \exp(-\xi/kT)} \tag{4.4}$$

where N_c is the volume density of states in the conduction band (for minority electrons) and ξ is the energy separation of the surface state and the conduction band edge (Carnes and Kosonocky, 1972; Lamb *et al*, 1974). For $N_c \simeq 10^{25}$ m^{-3} over the lowest kT of the conduction band and $\xi = 0.2$ eV, at room temperature we have

$$\tau_{\text{empty}} \simeq 30 \times 10^{-9}\ \text{s} \tag{4.5}$$

τ_{empty} increases considerably for surface states deeper in the forbidden energy gap (~10^{-4} s for mid band gap) but even equation 4.5 indicates a transfer period of about 120 ns (four time constants) for a residual charge as large as 2%, so that poor charge-transfer efficiency may be expected at clock frequencies of about 1 MHz if the surface state density is comparable with the mobile charge density.

4.2.2 Charge offset (fat zero)

The use of a charge offset to reduce charge-transfer inefficiency has already been mentioned in sections 2.12 and 3.7. The relatively rapid filling time of surface states when compared with their emptying time also allows the offset technique to be used to reduce considerably the charge smearing effect of surface states. In simple form the idea is to make the minimum charge transferred significantly greater than that which is released at each transfer so that the surface states are filled at all times when signal charge is present. Typically a zero level of 10% to 25% of saturation charge is required.

Two effects reduce the effectiveness of the charge offset. Lateral confinement of the charge-transfer channel by channel stop diffusions (section 2.3 and Fig. 2.8) causes a sloping potential profile which prevents the offset charge from filling some surface states that come into contact with the signal charge. This effect is illustrated in Fig. 4.2 which shows the lateral potential profiles for no contained charge, for an offset charge and for an offset charge plus signal charge.

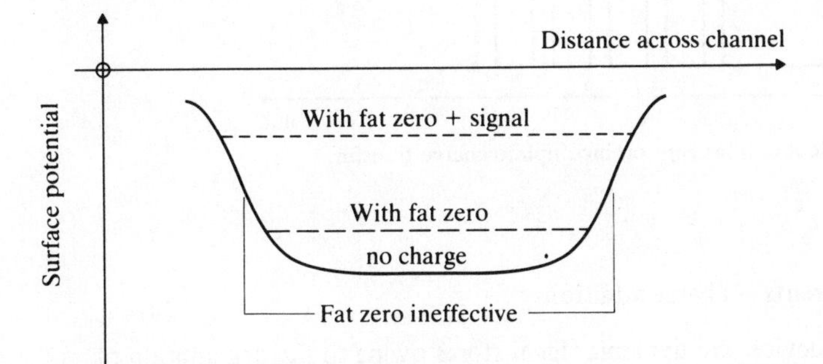

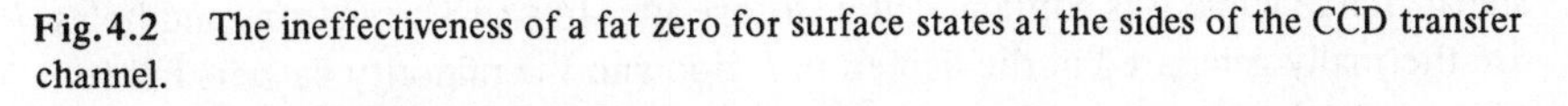
Fig.4.2 The ineffectiveness of a fat zero for surface states at the sides of the CCD transfer channel.

The second effect is caused by the existence of surface states with small capture cross-sections close to the band edge (Lamb *et al*, 1974). Equation 4.1 shows that these states have a longer filling time. If it is comparable with, or longer than, the charge containment period the surface states cannot be completely filled. The amount of trapped charge and its residual after the next charge-transfer event will then depend on the signal level.

4.2.3 Buried channel CCD

If the fat zero technique is unsatisfactory, it may be necessary to avoid the surface states by making charge transfer occur in the bulk of the semiconductor.

The buried channel CCD described in section 2.6 achieves this objective. Bulk trapping centres still exist but their concentration may be as low as 10^{15} m^{-3} so that the effects are much less serious. A small background charge must still be circulated continuously through the CCD if charge-transfer efficiencies approaching 10^{-5} per transfer are to be achieved (Mohsen and Tompsett, 1974). Fig. 4.3 illustrates the effect of a fat zero on the output signal of a CCD for an input signal of several 'ones' between many zeros.

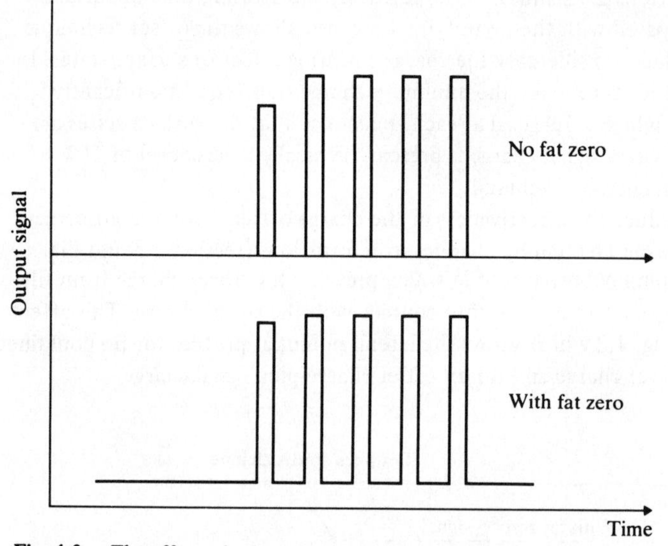

Fig. 4.3 The effect of a fat zero on incomplete charge transfer.

4.3 Leakage currents – charge addition

Charge-transfer devices are dynamic signal stores owing to the degradation of signal charge while it is contained at a storage site. In a CCD, electrons and holes are thermally generated in the depletion region and the minority carriers fall to the potential minimum while majority carriers move into the substrate. This action is just the same as that causing the reverse current in a p–n junction diode. There is also a contribution to the reverse current from diffusion in the majority carrier gradient but this is usually negligible in a silicon device (Sze, 1969). During the time a signal passes through the CCD, the total generated minority charge must be significantly less than any signal level if degradation is not to occur. This sets a lower limit on the clock frequency. The generation current density is J_{gen} given by

$$J_{gen} = n_i e l/\tau_r \qquad 4.6$$

where n_i is the intrinsic carrier density of the semiconductor under equilibrium conditions, l is the width of the depletion layer and τ_r is the lifetime of the minority carriers (Sze, 1969). Even though τ_r is a characteristic of a recombination event it appears in the above expression for a generation process because the two must be equal in an equilibrium semiconductor. For ideally pure silicon with electron–hole recombination directly across the band gap, τ_r is calculated to be $\simeq 1$ s (McKelvey, 1966). In real silicon, recombination and generation occur much more frequently via spurious energy levels in the band gap as described by the Shockley–Read–Hall mechanism (McKelvey, 1966). In this realistic case τ_r is typically as short as 10^{-3} s. At room temperature n_i in Si is $\sim 1.6 \times 10^{16}\ \text{m}^{-3}$ (Sze, 1969) so from equation 4.6 a typical generation current in a device with a 5 μm thick depletion layer is

$$J_{gen} = 1.3 \times 10^{-5}\ \text{A m}^{-2} \qquad 4.7$$

In a CCD with a cell size of 10 μm x 100 μm and a saturation charge of 1 pC the well will be filled by minority carrier generation in ~80 s. A better assessment is to consider the minimum clock frequency allowable for a CCD with one hundred storage elements to give an overall minority carrier injection of less than 1% of the saturation charge. The total transfer time must then be less than 0.8 s so that the clock frequency must be greater than 125 Hz.

In addition to the above generation currents there will be similar effects from leakage through the oxide by pinholes or similar defects introduced in the manufacture and there will be surface leakage currents at the semiconductor–oxide interface. The latter are similar to the bulk generation currents described above and have a higher current density (but flow through a smaller area) owing to the larger density of surface state defects decreasing the minority carrier lifetime. A further leakage current component can arise if a signal processing device is not completely screened from optical radiation, owing to the generation of electron–hole pairs by photons. Overall, these defects may easily increase the lower limit of the clock frequency by an order of magnitude. If particularly low leakage rates are required it is necessary to cool the device to reduce the thermal energy in the above processes. In equation 4.6 this would cause a decrease of n_i and therefore also of J_{gen}.

In a CCD each well will have a different leakage current in general. The operation of the CCD allows a simple presentation of the leakage pattern if the clock is switched off for a short time and then switched on at a high frequency. Fig. 4.4 shows the signal output from such an experiment.

4.4 Charge loss and substrate bias

In section 3.5 the loss of signal charge by recombination through finite gain in a bipolar BB was considered. Ideally this situation cannot occur in a CCD owing to the spatial separation of majority and minority carriers but care has to be taken to control the majority carriers under any electrode during the time when the clock bias is not applied. If the depletion region is removed during this time, majority carriers flood into those parts of the device which normally contain minority carriers in the storage phase. Some of the majority carriers would become trapped in surface states or other spurious energy levels. When the storage phase

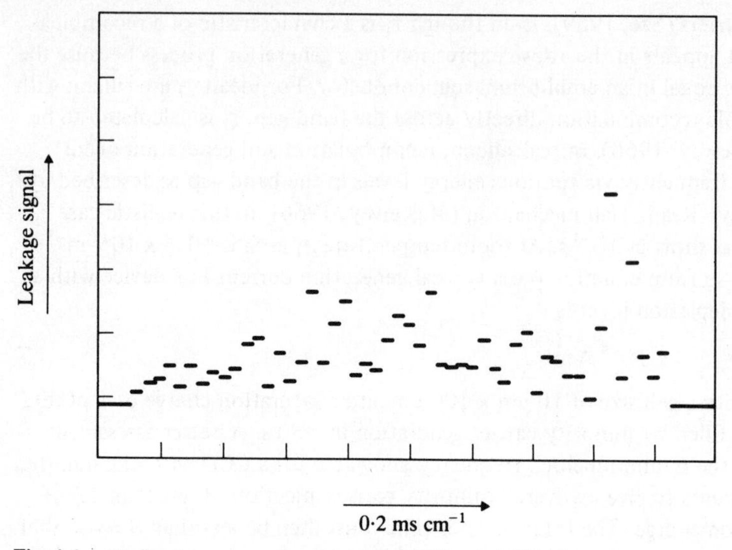

Fig. 4.4 The leakage current pattern of a CCD.

commences the free majority carriers are swept out of the storage region and minority carriers are transferred into it. Recombination can now occur by the Shockley–Read–Hall mechanism causing permanent loss of signal charge.

This signal loss may be avoided by maintaining the minimum clock voltage a little above zero potential so that the depletion region and inversion at the surface of a surface channel CCD, or the reverse bias in a buried channel CCD, are maintained at all times. Usually it is more convenient to apply a d.c. bias to the substrate of opposite polarity to the clock potential so that the clock design is simplified.

4.5 Potential humps and hollows

In section 2.11, potential humps were found to occur owing to geometrical factors in the electrostatic field distribution. Spurious charge stored on and in the oxide insulator between the electrodes and sometimes referred to as slow states can also cause humps or hollows in the surface potential of a surface channel CCD. This is illustrated in Fig. 4.5 for an n-channel device. The two neighbouring electrodes are held at the same positive potential. If the area density of the spurious charge, indicated by the circled symbols, is the same as the induced charge on the electrodes, the electric field lines will everywhere be perpendicular to the interfaces. For a higher electrode potential than this critical one the surface potential energy will be less negative under the gap than that under the electrodes, so that a potential hump occurs in much the same way as discussed in section 2.11. The electric field distribution is illustrated in Fig. 4.5(a). Similarly for a low electrode potential there will be a divergence of the electric field lines from the

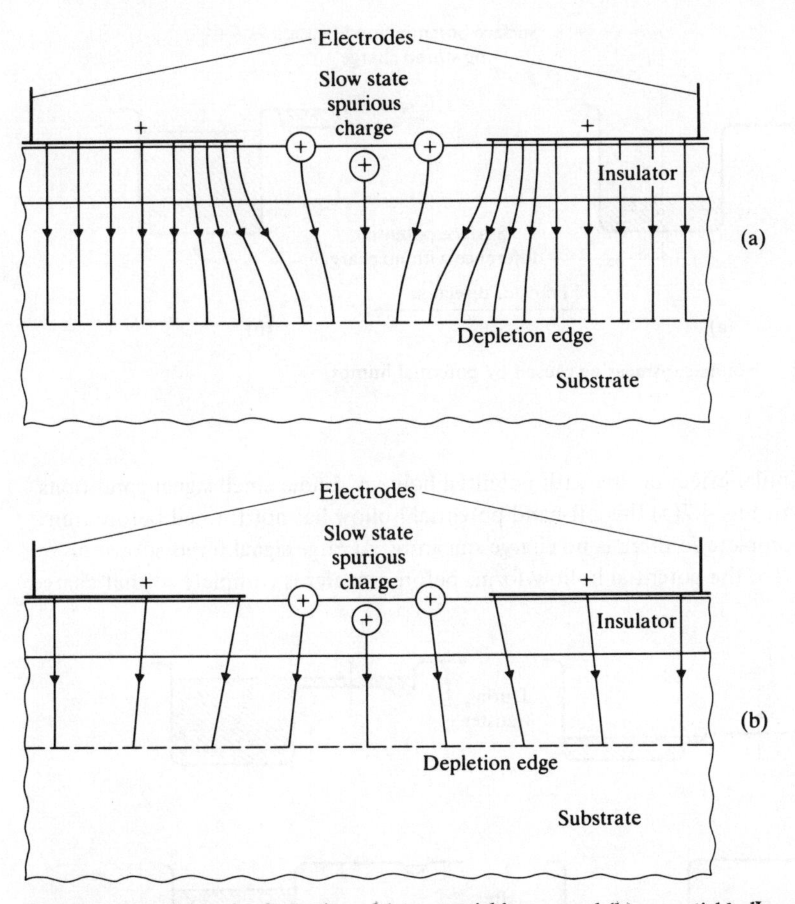

Fig. 4.5 Conditions for formation of (a) potential humps and (b) potential hollows from slow state charge between electrodes.

spurious charge so that a potential hollow exists under the gap. If the spurious charge had been negative a potential hump would have existed at all positive electrode potentials. For positive spurious charge the optimum clock bias necessary to avoid humps and hollows can be more conveniently provided by control of the substrate bias of opposite polarity.

Just as in section 2.11 the hump height or the hollow depth will vary from a maximum when the surface potential difference of neighbouring electrodes is zero down to zero when the surface potential difference is sufficiently large; this is a serious source of signal dependent charge smearing (Brodersen *et al*, 1975). The effect for potential humps is illustrated in Fig. 4.6. In Fig. 4.6(a) the stored charge is sufficiently small for the surface potential difference to remove the hump. In Fig. 4.6(b) the signal charge is large enough to raise the surface potential of the receiving well sufficiently to allow the existence of the potential hump so leaving a significant smeared charge to add to the following signal sample.

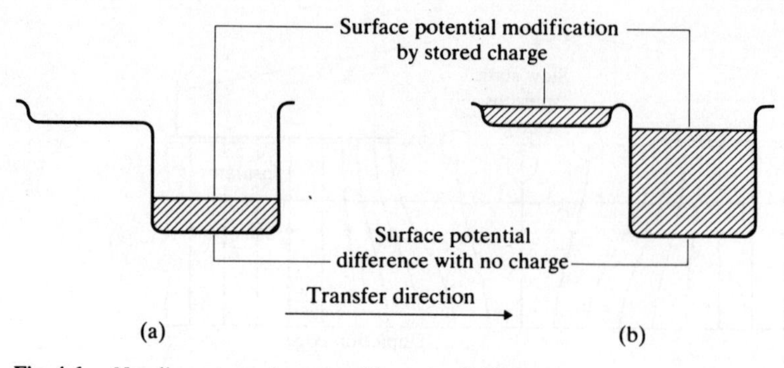

Fig. 4.6 Non-linear smearing caused by potential humps.

A similar effect occurs with potential hollows. Under small signal conditions shown in Fig. 4.7(a) the left-hand potential hollow has not formed before transfer is complete so there is no charge smearing. At large signal levels shown in Fig. 4.7(b) the potential hollow forms before transfer is complete so that charge

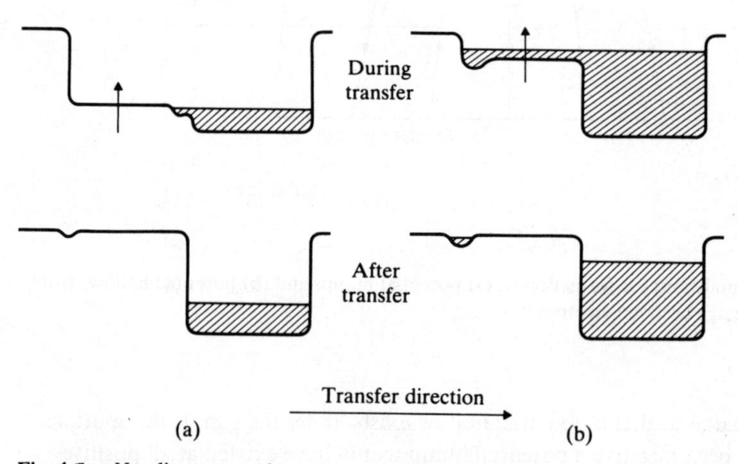

Fig. 4.7 Non-linear smearing caused by potential hollows.

smearing occurs. Charge smearing can also occur at small signal levels if the fall time of the clock waveform is so fast that the surface potential has risen to a point where hollows can occur before charge transfer is complete. The effect of both of these mechanisms is to give a non-linear charge smearing described by the fraction of charge left behind per transfer as illustrated in Fig. 4.8.

One particularly troublesome effect of the potential humps and hollows is their dependence on the slow-state spurious charge which may vary with the day to day ambient conditions. One way of avoiding this variability is to cover the surface insulation of the oxide insulator with a resistive 'sea' of polycrystalline

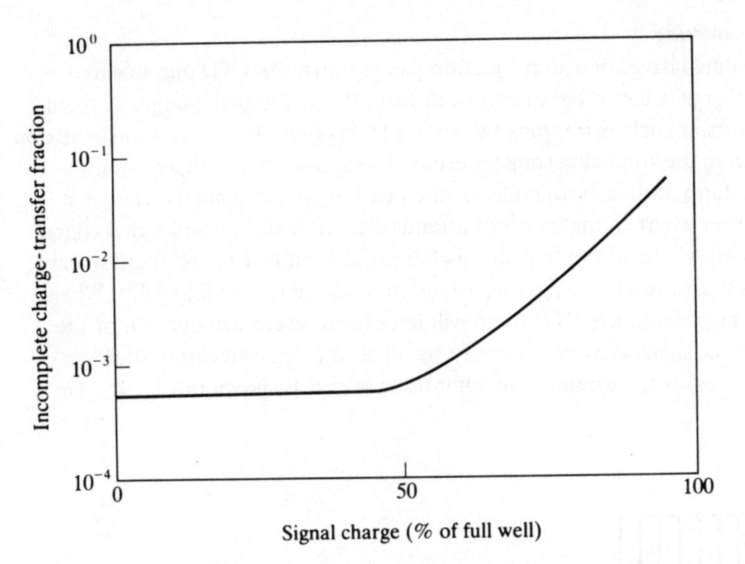

Fig. 4.8 Typical non-linear smearing behaviour caused by potential humps or hollows.

silicon which does not significantly affect the clock signals but does short out spurious charges. Details have been given by Kim and Snow (1972).

4.6 Signal distortion by charge smearing

The foregoing discussions of charge smearing have been concerned with its physical origins. In this section, the objective is to consider the distorting effect of charge smearing on some desirable signals and to introduce a convenient formulation for later descriptions of specific signal-processing devices. The various charge smearing mechanisms can be divided into three groups for convenience.

(i) Fixed smearing in which a constant amount of charge is lost per transfer. An example is surface state trapping loss.

(ii) Proportional smearing which is essentially the first-order approximation to those processes whose smearing is dependent on the amount of stored charge, such as the diffusion transfer in section 2.12 and the edge effect of surface-state trapping in section 4.2 and the finite gain limits for a BB in section 3.5.

(iii) Non-linear smearing where the linear proportional loss approximation is too poor to be of use as is the case with potential humps and hollows.

These effects can be measured in the time domain, either from the impulse response when the input is a single sample of charge, or from the step-function response on rising and falling edges when the input is a finite number of equal charges which must last for enough clock cycles to allow the steady state to be achieved. In both cases, the test signal is preceded and followed by a large number of signal zeros (or constant signals if a fat zero is used).

4.6.1 Fixed smearing

When the leading charge of a step function passes down the CTD one smeared quantity of charge is deposited in each well from the first signal charge. In fixed-smearing processes such as trapping by surface states there is usually some emission of charge during the following transfer event. The second signal charge refills these empty states and, in turn, collects an equal quantity of emitted charge at the next transfer event so that it is not attenuated. After the second signal charge reaches the well where all the first signal charge has been lost to the fixed-smearing processes it also suffers sequential attenuation in the succeeding wells. When the signal emerges from the CTD there will have been severe attenuation of the initial number of signal charges necessary to fill all the fixed-smearing sites. Beyond this number the attenuation will not be severe as shown in Fig. 4.9. On

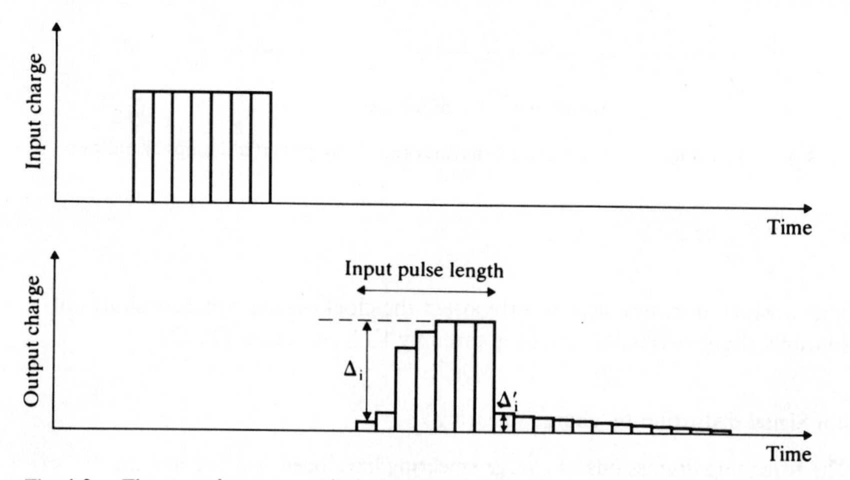

Fig. 4.9 The strongly asymmetrical charge smearing effects accompanying fixed smearing losses.

the falling edge the output signal continues at low level for many clock cycles until all the charge collected by fixed smearing has been emitted. Denoting the signal loss in each time interval after the leading edge by Δ_1, Δ_2 etc. and the emitted signal after the trailing edge by $\Delta_1{}^1$, $\Delta_2{}^1$ etc., the total fixed smearing in the CTD is *FL*, given by

$$FL = \left\{\sum_i \Delta_i\right\}_0 = \left\{\sum \Delta_1{}^1\right\}_0 \qquad 4.8$$

The zero subscript indicates other smearing processes are negligible. The considerable asymmetry of the output described by the differences of Δ_i and $\Delta_i{}^1$ is a characteristic feature of the fixed-smearing process. As mentioned in section 4.2, the fixed smearing is overcome by use of a fat zero which reduces the Δ_i considerably and maintains a continuous balance between fixed smearing and emission from it. Fig. 4.10 shows the non-linearity of a typical relationship between

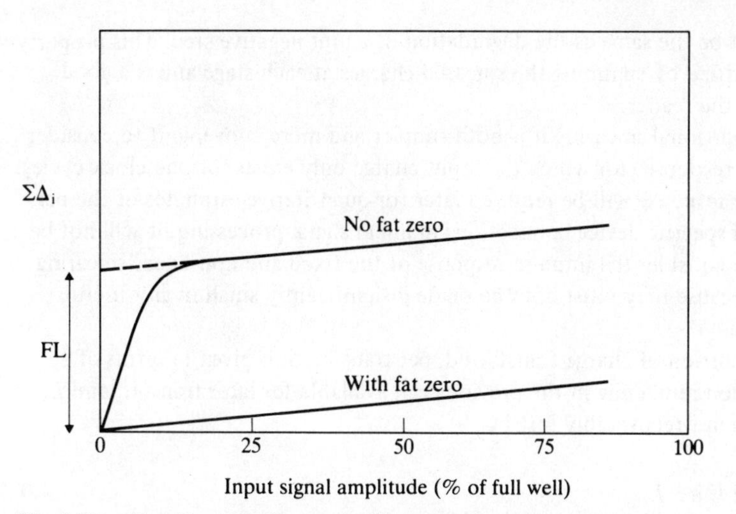

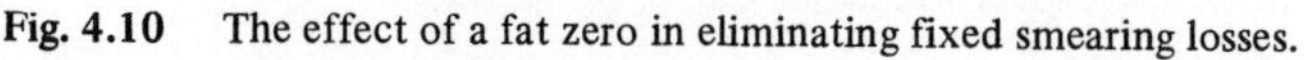

Fig. 4.10 The effect of a fat zero in eliminating fixed smearing losses.

the leading edge fixed smearing $\sum_i \Delta_i$ and the amplitude of the step-function input. There is a rapid rise as the amplitude is increased from zero until the total required fixed smearing loss is satisfied. If fixed smearing alone occurred, the curve would become horizontal beyond this point and the finite upward slope is caused by the proportional smearing considered next. When a sufficiently large fat zero is applied only the proportional smearing occurs.

4.6.2 Proportional smearing

When the charge left behind on transfer is directly proportional to the charge initially available for transfer the outputs corresponding to the positive and negative steps of the step function response must have the symmetry shown in Fig. 4.11 with $\Delta_i = \Delta_i^1$. That is, the degradation of a unit positive step in a linear

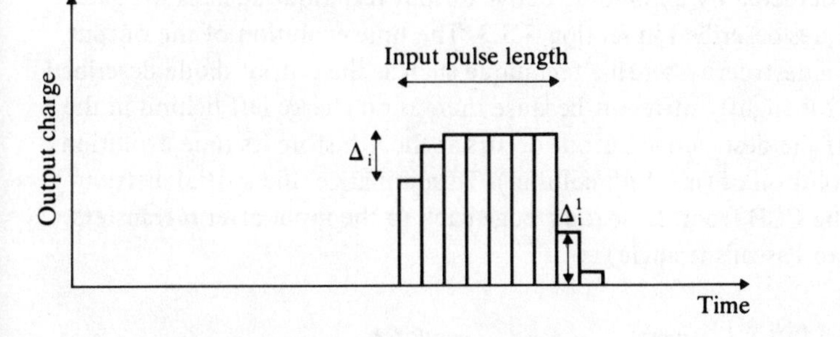

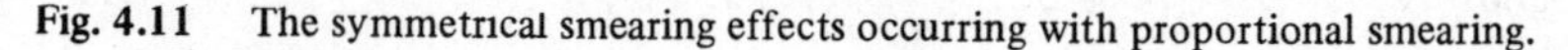

Fig. 4.11 The symmetrical smearing effects occurring with proportional smearing.

system must be the same as the degradation of a unit negative step. This property could be verified by summing the smeared charges at each stage and is a good exercise for the reader.

For proportional smearing it is both simpler and more convenient to consider the impulse response (for which the input charge only exists for one clock cycle). The impulse response will be required later for quantitative estimates of the performance of specific device connections in linear signal processing. It will not be necessary to consider the impulse response of the fixed and non-linear smearing processes because they must both be made insignificantly small in any useful linear device.

The proportion of charge transferred, per transfer, δ, is given in terms of ϵ, the proportion remaining in the previous cell available for later transfer, and L, the proportion irrecoverably lost by

$$\delta = 1 - \epsilon - L \tag{4.9}$$

In the general case of non-linear smearing

$$\epsilon = \Delta q_{\mathrm{r}}/\Delta q_0 \tag{4.10}$$

where Δq_{r} is an increment of the charge left behind and Δq_0 is the corresponding increment of the total charge involved in transfer. This definition is relevant for a small signal charge superimposed on a large and constant background charge q_0. In the following analysis it will be assumed that δ, ϵ and L are the same for all wells. L will usually be zero except for a bipolar BB.

During each transfer event some of the charge is transferred and some is not. No other choices are available but in the latter case there is also one extra unit of time delay equal to the (transfer period) x (number of clock phases). For a single unit of charge placed in one well at the input the charge becomes distributed between each storage unit in the same fashion that a Pascal's triangle of binominal coefficients is built up (Fig. 4.12). In this diagram a horizontal row describes the charge in each storage unit of the CTD at the appropriate transfer period after input of the impulse. Each column describes the time evolution of charge at one point of the CTD at intervals of one transfer period. It is essentially the charge that would be detected by a non-destructive output technique such as the gate sensing techniques described in section 5.3.3. The time evolution of the output signal through a destructive sensing technique such as the output diode described in section 5.3.2 is slightly different because there is no charge left behind in the output store. If the destructive output occurs at the nth store its time evolution is δ x (time evolution of $(n-1)$th column). To summarize, the spatial distribution through the CCD from the leading edge back to the input after n transfers (reverse order to Pascal's triangle) is

$$\delta^n, n\delta^{n-1}\epsilon, \frac{n(n-1)}{2!}\delta^{n-2}\epsilon^2, \ldots \frac{1}{(n-r)!\, r!}\delta^{n-r}\epsilon^r, \ldots \tag{4.11}$$

The output sequence for non-destructive sensing after r transfers is

$$\delta^r, (r+1)\delta^r\epsilon, \frac{(r+2)(r+1)}{2!}\delta^r\epsilon^2, \ldots \frac{(r+m)!}{r!\,m!}\delta^r\epsilon^m, \ldots \qquad 4.12$$

The output sequence for destructive sensing after p transfers is

$$\delta^p, p\delta^p\epsilon, \frac{(p+1)p}{2!}\delta^p\epsilon^2, \ldots \frac{(p+m-1)!}{(p-1)!\,m!}\delta^p\epsilon^n, \ldots \qquad 4.13$$

The frequency response (transfer function) for proportional loss is of importance in pointing out further problems arising from incomplete charge transfer.

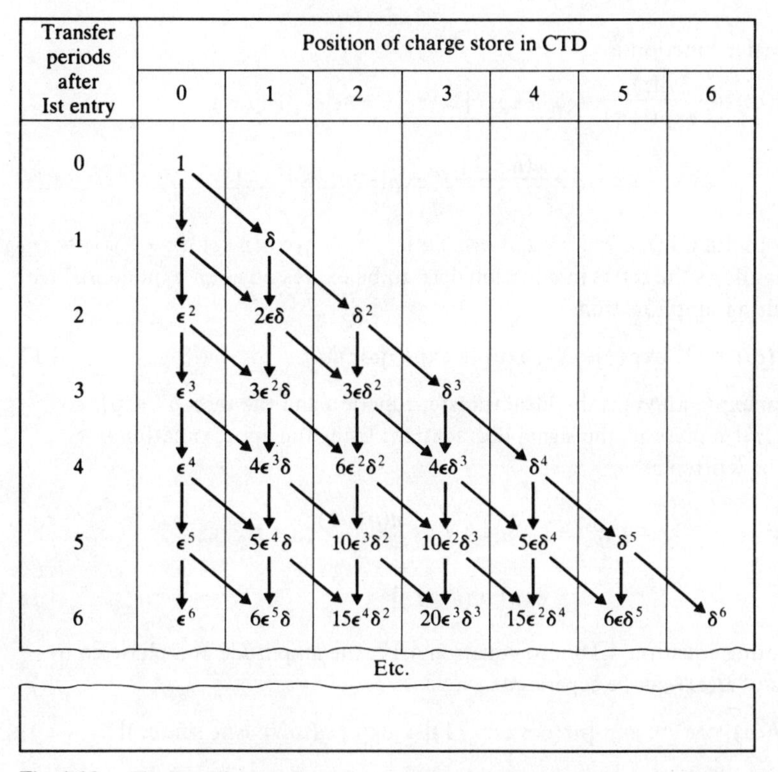

Fig. 4.12 The Pascal's triangle representation of charge-smearing coefficients.

It may be obtained 'mechanically' from the Fourier transform of equations 4.12 or 4.13 but it is instructive to build it up piecemeal as follows. The transfer function is simply the ratio of the sinusoidal output and input signals when a sinusoidal input is represented in the complex form $\exp(j\omega t)$. At any time t the output signal is the sum of all charge increments which have reached the output stage. For a destructive output stage described by equation 4.13 the output, $q(t)$, for an input of unit amplitude is

$$q(t) = \delta^n \times [\text{charge input at } (t - N\tau)]$$
$$+ n\epsilon\delta^n \times \{\text{charge input at } [t - (N+1)\tau]\}$$
$$+ \frac{n(n+1)\epsilon^2}{2} \delta^n \times \{\text{charge input at } [t - (N+2)]\ \tau\} + \ldots \qquad 4.14$$

where N is the number of stages in the CTD and $pN = n$ where p is the number of clock phases. τ is the delay per stage and is equal to p transfer periods.

i.e. $$q(t) = \delta^n \exp \mathrm{j}\omega(t - N\tau) + n\epsilon\delta^n \exp \mathrm{j}\omega[t - (N+1)\tau]$$
$$+ \frac{n(n+1)}{2} \epsilon^2\delta^n \exp \mathrm{j}\omega[t - (N+2)\tau] + \ldots \qquad 4.15$$

The transfer function is

$$H(\omega) = \frac{q(t)}{\exp(\mathrm{j}\omega t)} = \delta^n \exp(-\mathrm{j}\omega N\tau)[1 + n\epsilon \exp(-\mathrm{j}\omega\tau)$$
$$+ \frac{n(n+1)}{2} \epsilon^2 \exp(-2\mathrm{j}\omega\tau) + \ldots] \qquad 4.16$$

For any useful CTD, $n \gg 1$ so that $n(n+1)\epsilon^2 \simeq (n\epsilon)^2, n(n+1)(n+2)\epsilon^3 \simeq (n\epsilon)^3$ etc. This allows the series in equation 4.16 to be expressed as an exponential with the resulting simplification

$$H(\omega) = \delta^n \exp(-\mathrm{j}\omega N\tau) \exp[n\epsilon \exp(-\mathrm{j}\omega\tau)] \qquad 4.17$$

The term $\exp(-\mathrm{j}\omega N\tau)$ is the ideal transfer function and the term $\delta^n \exp[n\epsilon \exp(-\mathrm{j}\omega\tau)]$ represents the signal degradation. Using the approximation $n \gg 1$, δ^n, can be written as

$$\delta^n = [1 - (\epsilon + L)]^n = 1 - n(\epsilon + L) + \frac{n(n-1)}{2!}(\epsilon + L)^2 + \ldots$$
$$\simeq \exp[-n(\epsilon + L)] \qquad 4.18$$

Substituting equation 4.18 into equation 4.17, the amplitude and phase components of $H(\omega)$ can be separated.

i.e. $$H(\omega) = \mathrm{e}^{-nL} \exp\ [n\epsilon(\cos \omega\tau - 1)] \times \exp\ [-\mathrm{j}(\omega N\tau + n\epsilon \sin \omega\tau)] \qquad 4.19$$

The attenuation is

$$A(\omega) = \mathrm{e}^{-nL} \exp[n\epsilon(\cos \omega\tau - 1)] \qquad 4.20$$

At low frequencies the attenuation is only caused by charge loss but the charge smearing causes additional attenuation at high frequencies with a maximum at the Nyquist frequency when $\omega\tau = \pi$. In essence this is caused by the antiphase relationship of neighbouring output samples at the Nyquist frequency so that any smeared signal is directly subtracted from the following sample.

Charge smearing also causes a phase error owing to the additional component of signal delay that it introduces. The phase angle error relative to the linear phase shift $e^{-j\omega N\tau}$ is

$$\Delta\beta(\omega) = n\epsilon \sin \omega\tau \tag{4.21}$$

It is zero at low frequencies and also at the Nyquist frequency where the smeared components of the output, as remarked above, have phase shift of multiples of π so that they are only subtracted from the main signal without producing phase shift. The maximum phase error occurs at half the Nyquist frequency when $\omega\tau = \pi/2$ because the smeared components have their maximum phase difference (multiples of $\pi/2$) from the main signal. $A(\omega)$ and $\Delta\beta(\omega)$ for a large $n\epsilon$ of 0.25 and $L = 0$ are illustrated in Fig. 4.13. In a real device the amplitude transfer function will suffer further modification by the transfer function

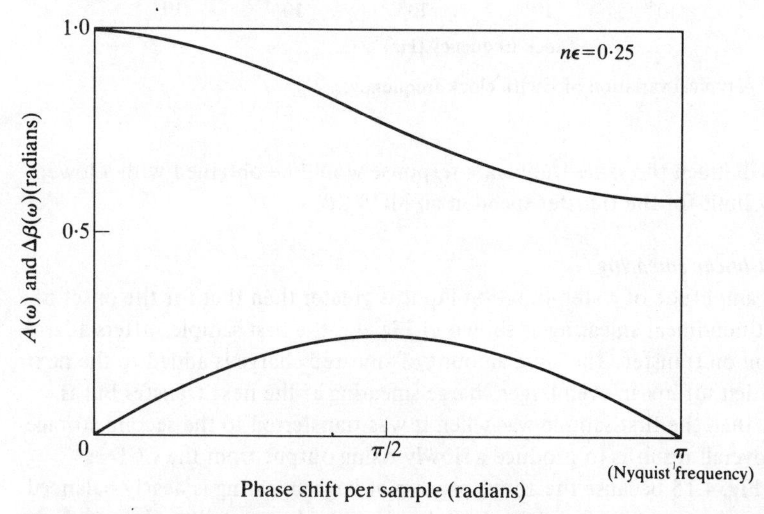

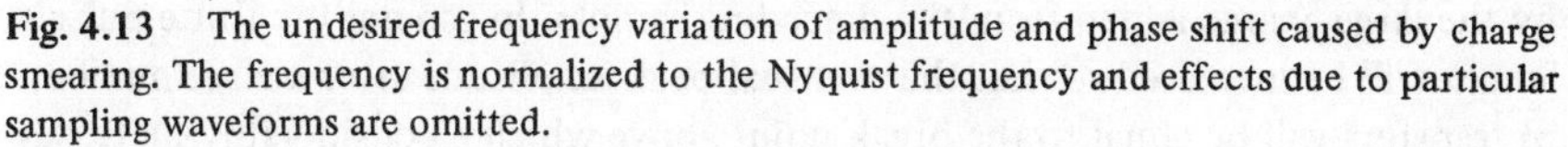

Fig. 4.13 The undesired frequency variation of amplitude and phase shift caused by charge smearing. The frequency is normalized to the Nyquist frequency and effects due to particular sampling waveforms are omitted.

associated with the sampled form of the output signal. The sampled nature of the signal has only been partially taken into account here by recognizing that the smeared signals occur at fixed intervals relative to the desirable signal. The extra modification will have a $(\sin x)/x$ form if the signals have a constant level between sampling points. Further details are given in Chapter 5.

Implicitly the above analysis was only valid for a single clock frequency owing to the dependence of ϵ on the transfer time. Owing to the several sources of incomplete charge transfer and their complicated dependence on particular details of the device it is not possible to give a general frequency dependence of ϵ. However, the behaviour illustrated in Fig. 4.14 is typical for a surface channel CCD. For a buried channel CCD the good transfer efficiency would extend to higher frequencies and the low frequency ϵ would be smaller.

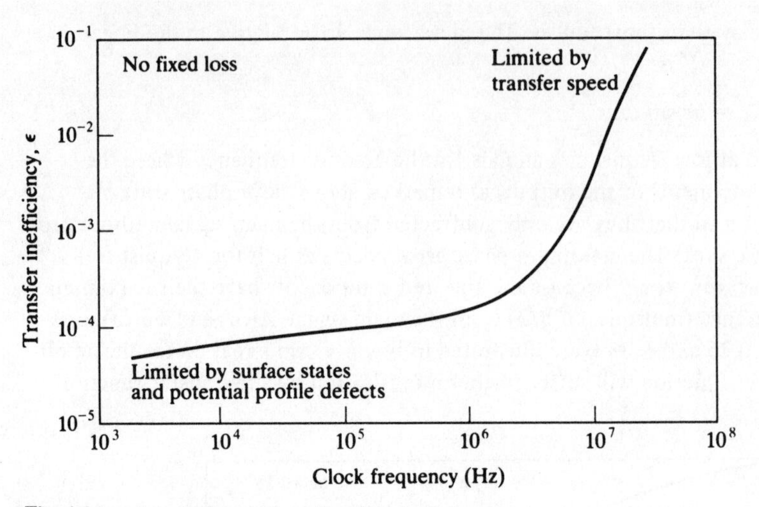

Fig. 4.14 A typical variation of ϵ with clock frequency.

For a BB much the same frequency response would be obtained with a lower frequency limit for the transfer speed in an MOS BB.

4.6.3 Non-linear smearing

When the amplitude of a step-function input is greater than that for the onset of significant non-linear smearing as shown in Fig. 4.8 the first sample suffers a large attenuation on transfer. The large amount of smeared charge is added to the next sample which suffers an even larger charge smearing at the next transfer but is still larger than the first sample was when it was transferred to the second storage cell. The overall result is to produce a slowly rising output from the CCD as shown in Fig. 4.15 because the above process of large smearing is nearly balanced by the large charge added from the preceding sample. In general the finite pulse length will be increased so that the eventual peak amplitude after a large number of transfers will be equal to the break point above which ϵ considerably increases

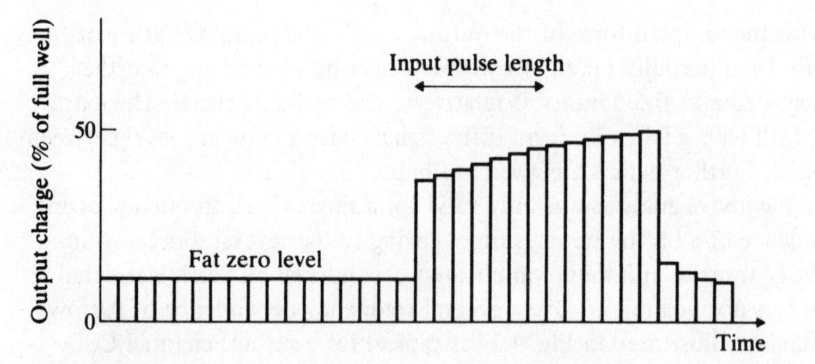

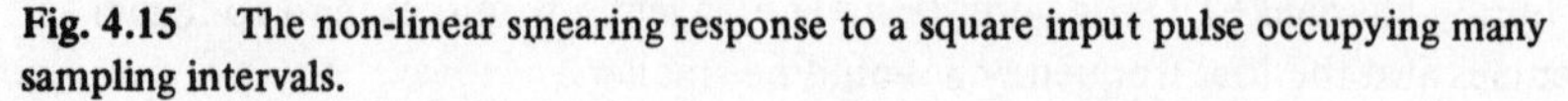

Fig. 4.15 The non-linear smearing response to a square input pulse occupying many sampling intervals.

(Fig. 4.8). There is a steep reduction of the amplitude following the last unit of smeared charge which is large enough to experience the excessive smearing effect. Beyond this, ϵ only has its small, proportional-smearing value. This highly asymmetric charge smearing with slow rise and fast fall is characteristic of non-linear smearing. It has a serious distortion effect on a sine wave as shown in Fig. 4.16, the output for signal extremes at 10% and 95% of the full well. The device

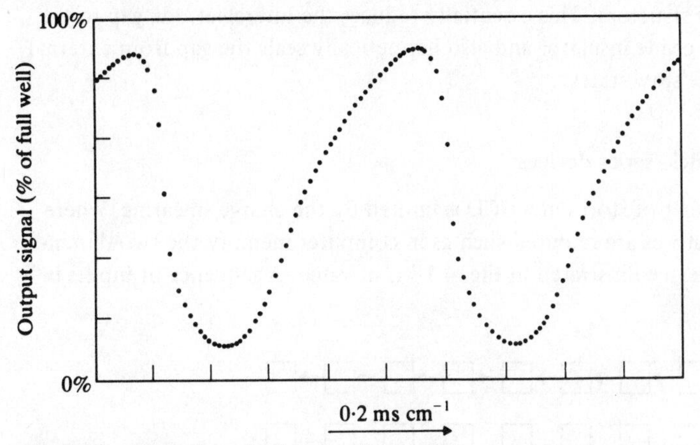

Fig. 4.16 Distortion of a sine wave by non-linear smearing.

was a 100 element, three-phase CCD with 2 μm gaps between its metal electrodes. The effect can be largely removed by using an alternate input sampling technique in which alternate wells are left empty at the input. In this way the large charge smearing occurs into empty wells so that adjacent, alternate input signals are only smeared by proportional smearing. The output from the wells into which the input charge was placed has the clipped sine wave form shown in Fig. 4.17

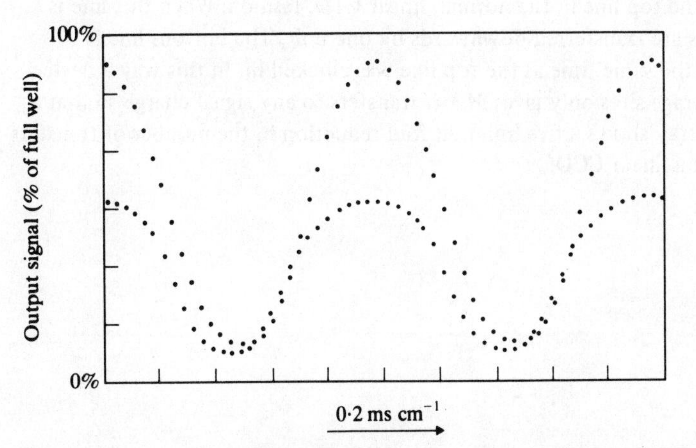

Fig. 4.17 Correction of the distortion in Fig. 4.16 by the alternate input technique.

but summation of this signal and the smeared signal gives a high quality reproduction of the sine wave input. A disadvantage of this technique arises from the doubling of the number of CCD stages required to handle the same number of signal samples but the overall signal improvement is still considerable. Quantitative details have been given by Tozer and Hobson (1976a).

A better device design which considerably reduces or completely removes the potential hump/hollow problem and highly non-linear smearing is the polysilicon overlapping-gate structure. This essentially reduces the interelectrode gap to the thickness of the oxide insulator and also hermetically seals the gap from external influences on the slow states.

4.7 Series–parallel–series devices

The overall number of stores in a CCD is limited by the charge smearing. Where extremely long stores are required such as in computer memory the two-dimensional transfer device illustrated in Fig. 4.18 is of value. A sequence of inputs is

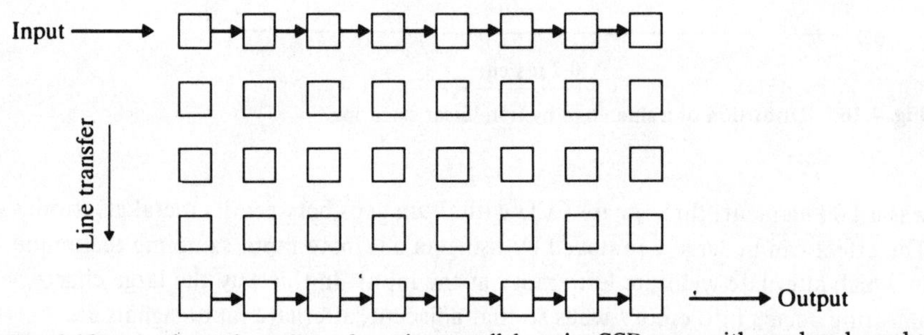

Fig. 4.18 A schematic outline of a series–parallel–series CCD array with a reduced number of transfer events between input and output.

introduced in the top line in the normal, linear CTD, fashion. When this line is full all the lines are transferred downwards by one unit. The bottom line is clocked out at the same time as the top line was clocked in. In this way a device with $M \times N$ storage sites only gives $M + N$ transfers to any signal charge so that a 1000×1000 array shows a five hundred fold reduction in the number of transfers compared with a linear CCD.

5

CCD Circuit Interfacing and Noise Considerations

5.1 Introduction

The subjects of noise and circuit interfacing of CCD naturally fall together owing to the small number of electrons available to carry signals.

The saturation charge of a single well in a CCD is typically $\sim 10^{-12}$ C which is $\sim 10^7$ electrons. Signal-processing applications may require a dynamic range of 60 dB or better and low-light optical applications are more demanding. There is a considerable problem in linearly converting external circuit voltages or currents to a charge whose magnitude is only reckoned in thousands of electrons and doing so in periods of no longer than 1 μs in many cases. Initially the problem of introducing the signal to the CCD will be dealt with and comments will later be applied to methods of linear extraction. This will be followed by noise considerations. Unless otherwise stated the descriptions will apply to surface channel CCD on a p-type substrate so that minority electrons carry the sampled signal.

5.2 Input circuits

5.2.1 FET input

The MOSFET form of the input circuit has been described in section 2.14. The input voltage can be applied to either the input diode or the input gate in a common gate or common source mode respectively but current will only flow when the clock voltage is applied to the ϕ_1 electrode. The charge injected is

$$q = \int_{\substack{\phi_1 \\ \text{clock} \\ \text{period}}} i \, \mathrm{d}t \tag{5.1}$$

$$\text{i.e.} \quad q = \gamma(V_G - V_{ID} - V_T)^2 \tau = \mu \frac{C_{ox} A}{2\omega_{G^2}} (V_G - V_{ID} - V_T)^2 \tau \tag{5.2}$$

An explicit form of γ (see Chapter 3) has been used in the parabolic relationship between MOSFET current and gate voltage V_G, (Sze, 1969) in equation 5.2. μ is the minority carrier mobility, C_{ox} is the oxide capacity per unit area under the input gate, ω_G is the input gate length, A is the input gate area, V_{ID} is the input diode voltage, V_T is the threshold voltage and τ is the transfer period defined by the clock period. The maximum stored charge or full-well charge is set by the size of the charge-transfer storage sites. If these sites have the same width as the input gate but have a length w the maximum input charge is $\sim(w/w_G)A\,C_{ox}\,V_c$ (see section 2.7) where V_c is the clock voltage amplitude. Therefore the working range of V_{ID} or V_G from zero injected charge, when $V_G - V_{ID} - V_T = 0$, to q_{sat} is given by

$$V_{I\,max} = \sqrt{\left(\frac{2w\,w_G\,V_c}{\mu\tau}\right)} \qquad 5.3$$

For $V_c = 15$ V, $w = w_G = 10\ \mu$m, $\mu = 0.05\ m^2 V^{-1} s^{-1}$ and $\tau = 10^{-6}$ s,

$$V_{I\,max} \simeq 0.25 \text{ V}$$

This input mode causes a large non-linearity owing to the parabolic relationship between charge and voltage given in equation 5.2. It is not used except in the most elementary testing. The charge injected is also dependent on the clock period but this can be avoided by pulsing the input circuit into conduction for a fixed duration using the electrode not required for the input signal. Typical distortion figures relative to the fundamental are 12 dB for second harmonic and 23 dB for third harmonic for an input signal equal to half the full well signal (Sequin and Mohsen, 1975).

5.2.2 Diode cut-off

A much more satisfactory input technique is illustrated in Fig. 5.1. The input signal voltage is applied to the input diode so that the surface potential of its

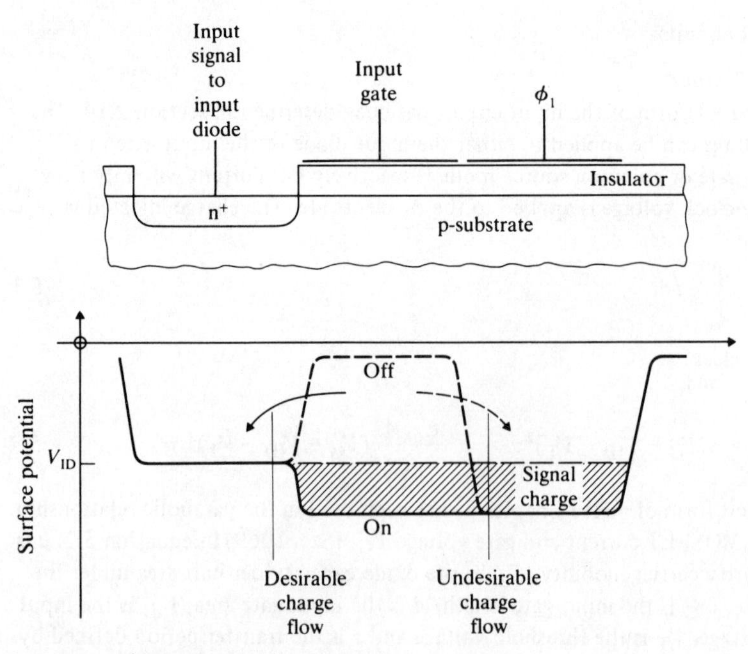

Fig. 5.1 The diode cut-off input technique.

electrons lies within the surface potential range of ϕ_1, for no stored charge, with and without the clock signal applied. Just before or during the time ϕ_1 is on, the input gate is switched to a positive voltage equal to the clock voltage, so that its surface potential falls to a value equal to that of ϕ_1 when the clock is on. Charge then flows under the input gate and under ϕ_1 when it is on. Before ϕ_1 goes off the input gate potential is removed so that its surface potential rises to the *off* value. The charge temporarily stored under the input gate then flows back into the limitless supply of electrons in the input diode region isolating under the ϕ_1 electrode a charge which is linearly related to the input diode voltage. A disadvantage of this technique is that the input gate potential must be removed slowly so that the charge under the input gate can all flow back into the input diode region. Rapid removal of this potential causes a flow of charge both to ϕ_1 (shown as undesirable charge flow in Fig. 5.1) and the input diode and causes a considerable increase in noise owing to the random nature of the charge partition (Carnes *et al*, 1973).

The linearity of this mode is considerably better than that of the FET mode. The surface potential associated with the injected charge is equal to the electron potential beneath the input diode which, in turn, is set by the input diode reverse voltage. The constant of proportionality between the injected charge and the surface potential is approximately the oxide capacity. The residual non-linearity is that discussed in section 2.7. In a buried channel CCD this source of non-linearity would be much greater, as illustrated in Fig. 2.21 owing to the variable position of the stored charge relative to the surface. However, the following source of non-linearity tends to compensate this one so that the overall linearity in a buried channel device is not as bad as may be expected.

The shape of the potential profile under the input gate causes additional non-linearity. In practice it is not perfectly square and an irregular peak will cause some partition of charge into the ϕ_1 region as the input gate voltage is removed. The magnitude of the irregularities will change with the magnitude of the input gate voltage (Sequin and Mohsen, 1975) owing to fringing field effects and simple scaling with the mean voltage. The partitioning effect will decrease with decreasing gate voltage at the cut-off point defined by the signal level on the input diode. Overall the typical second harmonic distortion of this mode in a surface channel CCD is 23 dB below the fundamental for an input signal equal to 50% of the full well saturation charge.

The ideal potential profile would have a much narrower 'knife edge' shape to decrease this partition non-linearity. The greater fringing field effects in a buried channel device cause rounding of the potential profile under the input gate so that the ideal condition is more closely approached. In addition, the residual partition non-linearity tends to compensate the charge position non-linearity illustrated in Fig. 2.21. The second harmonic distortion of a buried channel device in this mode is typically 30 dB below the fundamental and the third harmonic distortion is more than 40 dB below the fundamental for an input signal equal to 50% of full well.

The range of input voltages in this mode is equal to the clock-voltage amplitude.

5.2.3 Fill and spill

This technique has various names including charge preset, charge equilibration and potential equilibration. The surface potentials of an n-surface-channel CCD illustrate the mode of operation in Fig. 5.2. Throughout one charge injection process the input signal plus the necessary d.c. bias are essentially constant so that the surface potential under the input gate is constant. During the ϕ_1 clock

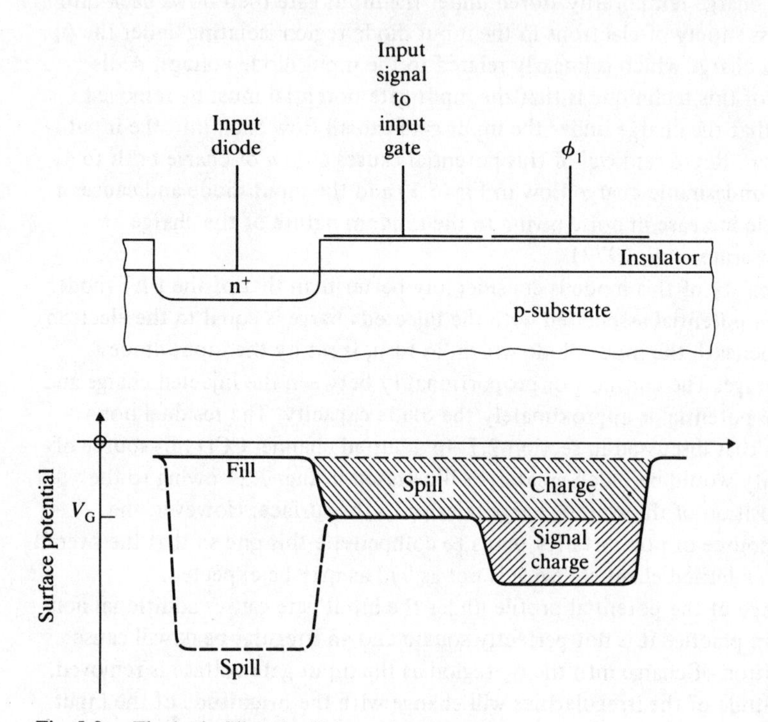

Fig. 5.2 The basic fill and spill input technique.

period the input diode is pulsed to a low voltage, causing a correspondingly high surface potential for the *fill* operation. Charge is injected from the infinite electron source of the input diode to raise the surface potential, under ϕ_1 and the input gate, to the *fill* level. Before ϕ_1 potential is removed, the input diode potential is allowed to return to its high quiescent *spill* voltage so that the corresponding surface potential falls to a value less than that under the ϕ_1 electrode. Excess charge then spills out into the input diode until the surface potential under ϕ_1 is equal to the surface potential defined without charge storage by the signal voltage under the input gate. The range of the input voltage in this mode is equal to the clock-voltage amplitude if all electrodes have the same area. The low input-diode voltage must be greater than the lowest value of the clock voltage so that its surface potential never becomes high enough to cause charge to flow 'over the top' of the potential barrier under the first ϕ_2 electrode.

The linearity of this input mode is better than that of the diode cut-off mode because the conversion of voltage to charge is not dependent on the capacity of the depletion region. The surface potential under ϕ_1 after the fill and spill operation is equal to the surface potential under the input gate. Accordingly, the difference of voltages, ΔV_{in}, on the input gate and ϕ_1 is taken up on the insulator. If $C_{ox}A$ is the capacity of the oxide insulator under ϕ_1 and the stored charge is Δq,

$$\Delta q = C_{ox}A \, \Delta V_{in} \tag{5.4}$$

The relationship between the injected charge and the input voltage has the form shown in Fig. 5.3. At low voltage the cut-off occurs when the *fill* operation can-

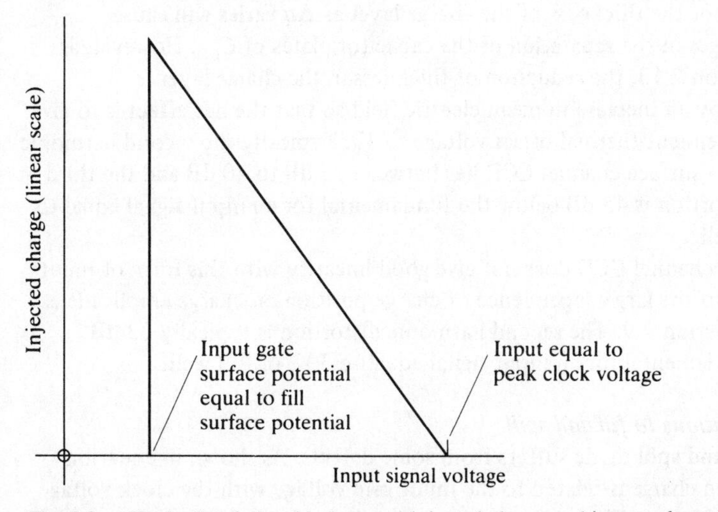

Fig. 5.3 The basic relationship between injected charge and input voltage for the fill and spill technique of Fig. 5.2.

not be carried out. In practice it is not a sharp cut-off owing to the finite time allowed for the fill operation and the restriction on the charge-transfer rate as the input gate surface potential approaches the input-diode surface potential (see section 2.12). This limit of the characteristic becomes steeper for longer fill times.

Equation 5.4 can be derived in a more formal way as follows. The surface potential, V_s, is a function of the electrode bias voltage, V_B, minus that part of the voltage drop across the oxide insulator $\Delta q/C_{ox}A$, which may be ascribed to the stored minority carriers.

i.e. $$V_s = f\left(V_B - \frac{\Delta q}{C_{ox}A}\right) \tag{5.5}$$

The fill and spill input causes potential equilibration between the surface potential under ϕ_1 (with bias voltage V_c) and the surface potential under the input gate (with bias voltage V_G and $\Delta q = 0$).

$$\therefore \quad f\left(V_c - \frac{\Delta q}{C_{ox}A}\right) = f(V_G) \tag{5.6}$$

$$\therefore \quad V_c - \frac{\Delta q}{C_{ox}A} = V_G$$

$$\text{i.e.} \quad \Delta V_{in} = V_c - V_G = \frac{\Delta q}{C_{ox}A} \tag{5.7}$$

Equation 5.7 is the same as equation 5.4.

For good linearity C_{ox} must not depend on Δq which requires, on first consideration, that Δq is stored in a layer which is significantly thinner than the oxide thickness. This is not necessarily a good approximation as shown in section 2.11. Changes of the thickness of the charge layer as Δq varies will cause effective changes in the separation of the capacitor plates of C_{ox}. However, as shown in section 2.13, the reduction of thickness in the charge layer is accompanied by an increase in mean electric field so that the net effect is to give a charge independent thermal offset voltage to V_c. Typically the second harmonic distortion for a surface channel CCD lies between 35 dB to 40 dB and the third harmonic distortion is 45 dB below the fundamental for an input signal equal to 50% of full well.

The buried channel CCD does not give good linearity with this form of input circuit owing to the large dependence of charge position on charge amplitude as discussed in section 2.9. The second harmonic distortion is typically 33 dB below the fundamental for an input signal equal to 50% of full well.

5.2.4 Modifications to fill and spill

The basic fill and spill mode suffers from some defects. As shown in equation 5.6 the injected charge is related to the input gate voltage with the clock voltage of ϕ_1 as the reference. This causes spurious signals on the clock lines to be directly mixed with the signal charge. If the first ϕ_1 electrode is isolated from all the other electrodes and has a d.c. bias applied which is equal to half the clock-voltage amplitude this defect is overcome. The fill and spill operation then takes place during the ϕ_1 period but the charge is received into an isolated well. When ϕ_2 is applied charge transfer takes place in the normal way as shown in Fig. 5.4. The action has similarities to the psuedo-two-phase operation described in section 2.5.3 and in both cases the signal handling capacity is reduced by a factor of two. However it can be restored in this case by making the second gate twice as long as the clock gates.

Fringing fields effects make the effective area of the metering electrode (G_2 in Fig. 5.4) depend on the input voltage. The effect will be greatest between electrodes G_2 and ϕ_2 in Fig. 5.4. During the input period a large potential difference will exist at this boundary for all but the largest charge inputs. The resulting potential gradient will have a finite slope so that the effective electrode area will be smaller for small charge input, corresponding to a high voltage on G_1, as shown in Fig. 5.5. The effect will occur to a lesser extent between G_1 and G_2 where the equilibration surface potentials are equal. At this point the fringing

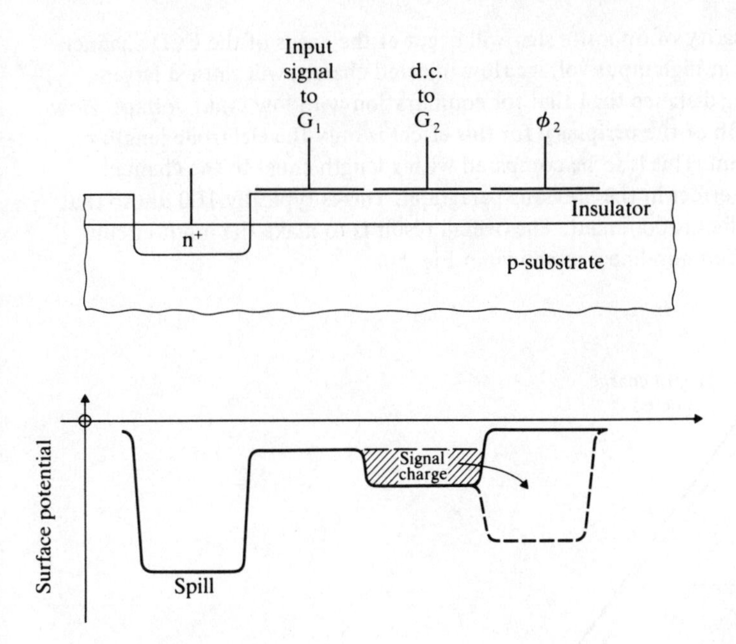

effects will arise from the difference of voltage across the oxide insulator under G_1 and G_2 and will only occur over a distance of the order of the oxide thickness. However an overall change of effective electrode length equal to the oxide thickness will cause a non-linearity of 1% (−40 dB) for an oxide thickness of 0.1 μm and an electrode length of 10 μm.

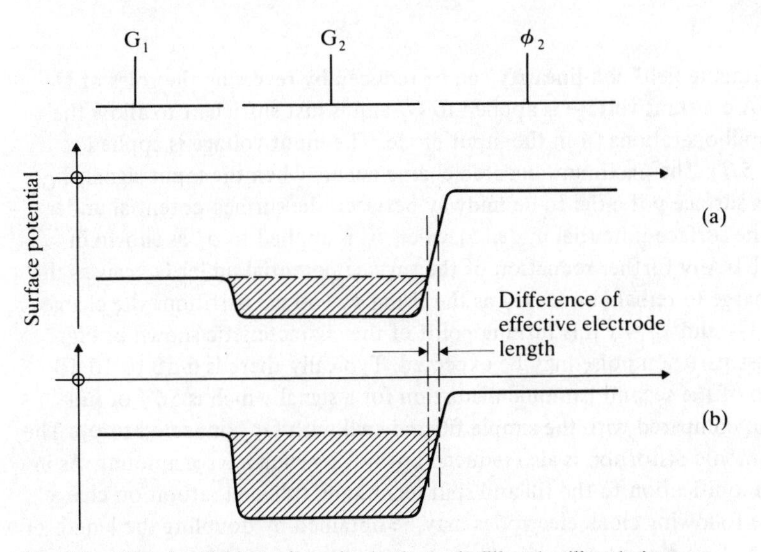

Fig. 5.5 Fringing-field non-linearity in the basic fill and spill technique.

A non-linearity of opposite sign will occur at the edges of the CCD channel. Equilibration at high input voltage (low injected charge) will cause a larger lateral fringing distance than that for equilibration with low input voltage. However, the length of the periphery for this effect is only the electrode length or typically 10 μm. This is to be compared with a length equal to the channel width for the effect in the previous paragraph. This is typically 100 μm so that the former effect is dominant. The overall result is to make the input circuit transfer function non-linear as shown in Fig. 5.6.

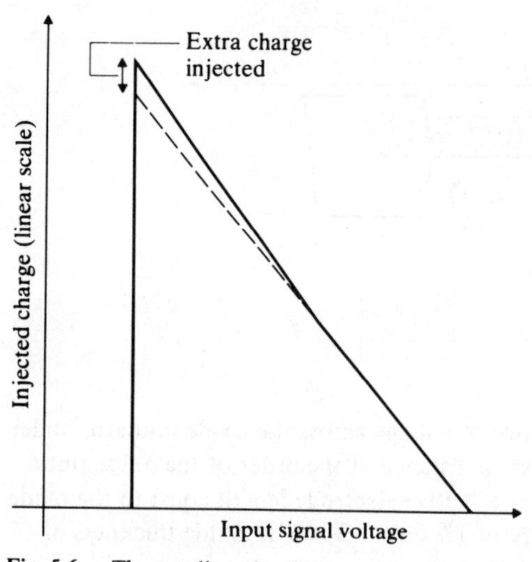

Fig. 5.6 The non-linearity illustrated by Fig. 5.5.

The fringing field non-linearity can be reduced by reversing the roles of G_1 and G_2. A constant voltage is applied to G_1 and is just sufficient to allow the fill and spill operations from the input diode. The input voltage is applied to G_2 (Fig. 5.7). The maximum injected charge occurs when the input signal, V_{G_2}, causes its surface potential to lie midway between the surface potential under G_1 and the surface potential under ϕ_2 when V_c is applied to ϕ_2 as shown in Fig. 5.7(b). Any further reduction of the surface potential under G_2 causes the excess charge to remain under G_2 as the rising ϕ_2 voltage partitions the charge between G_2 and ϕ_3. At this turning point of the characteristic shown in Fig. 5.8, excess partition noise may be expected. Typically there is 6 dB to 10 dB reduction of the second hamonic distortion for a signal which is 50% of full well when compared with the simple fill and spill with G_2 connected to ϕ_1. The third harmonic distortion is also reduced but not by such a large amount. As in the first modification to the fill and spill technique, full well saturation charge under the following clock electrodes may be obtained by doubling the length of G_2 and the first ϕ_2 and ϕ_3 electrodes.

One final advantage of the reversed use of G_1 and G_2 arises if they have different oxide thicknesses. Referring to equation 5.6 equilibration always occurs at a constant surface potential, V_{S1}, so that, with the appropriate subscripts,

$$V_{G_2} - \frac{\Delta q}{C_{OX}A} = V_{S1} = \text{constant} \qquad 5.8$$

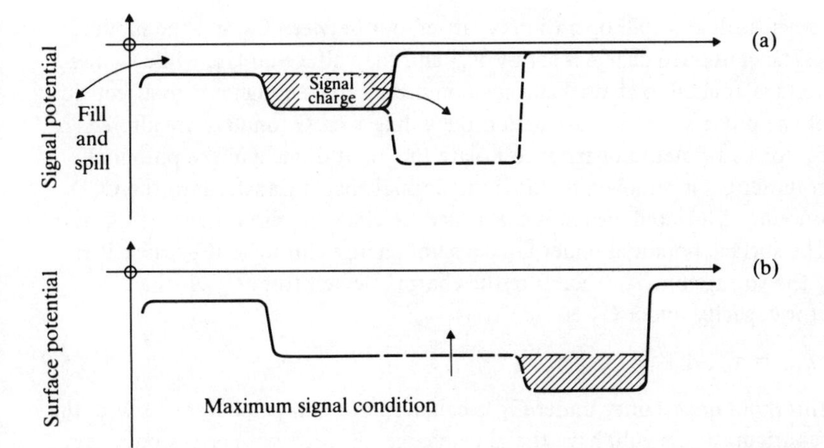

Fig. 5.7 A fill and spill technique which always equilibrates at the same surface potential.

Therefore Δq is linearly proportional to V_{G_2}. If G_1 and G_2 had different oxide thicknesses the functions on either side of equation 5.6 would be different so that the linear condition in equation 5.7 would not emerge.

One problem in the control of the input charge is the lack of tight control on the threshold voltage of each MOS element. In general this will cause an unknown offset voltage and a corresponding offset in the injected charge. This can be a serious problem in large scale integrated applications where many equal injections are required, such as the parallel fat zeros along each line of an imaging device. It may be avoided by using the same electrode, G_2, to control both the charge injected in a fill and spill operation and the amount of charge subsequently released, as shown in Fig. 5.9. Initially, with a low voltage on G_5 so that it forms a potential barrier at the entrance to the main CCD transfer

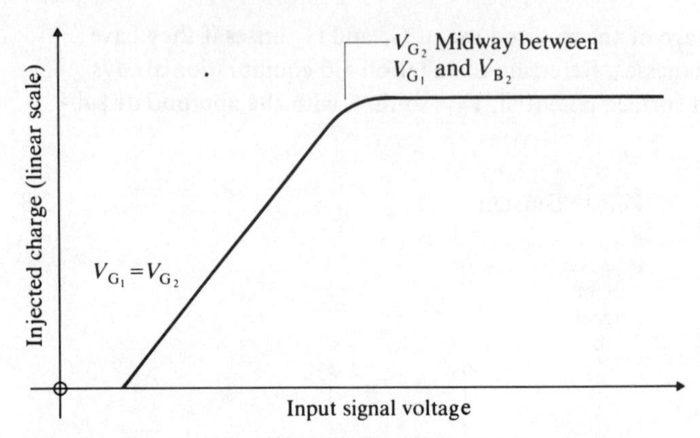

Fig. 5.8 The relationship between injected charge and input voltage for the input technique of Fig. 5.7.

electrodes, a fill and spill operation is carried out between G_2 and the input diode. The contained charge is set by V_{ref} and the voltage on G_3, while G_1 has a high voltage applied so that its surface potential is low enough not to affect this part of the process. In the next period the voltages on G_1 and G_5 are altered so that G_1 forms a potential barrier to isolate the input diode and the potential barrier under G_5 is removed to allow the normal charge transfer into the CCD. The amount of injected charge is controlled by changing the voltage on G_2 by V_{in}. The surface potential under G_2 does not change throughout because it is set by the voltage on G_3. Therefore the charge injected from $C_{ox}A$, the insulator capacity under G_2 is

$$q_{in} = C_{ox}A\ V_{in} \qquad 5.9$$

Any threshold uncertainty under G_2 is balanced out and parallel tracks with the same configuration would have the same charge injection with errors only dictated by the differences in areas of corresponding electrodes. In an

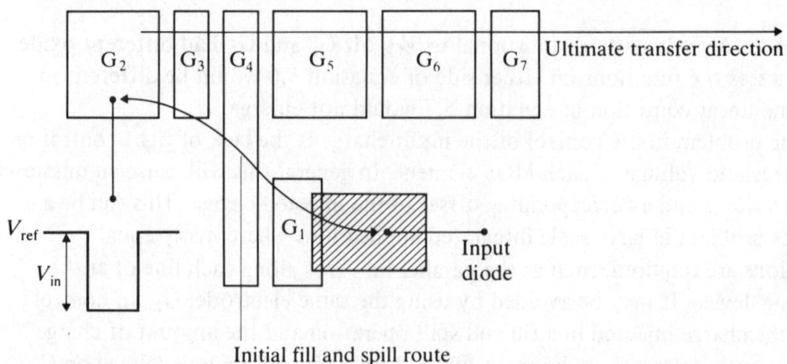

Fig. 5.9 A threshold-insensitive fill and spill input technique.

alternative arrangement, the input signal, of opposite polarity, could be applied to G_3. Both versions of this input method suffer from the disadvantage that they are not self-sampling and require the input signal to be provided in the form of a well-defined pulse.

5.2.5 Capacitive metering

The linearity of the input transducer depends on proportional conversion of voltage to charge. The capacitive metering technique illustrated in Fig. 5.10 uses

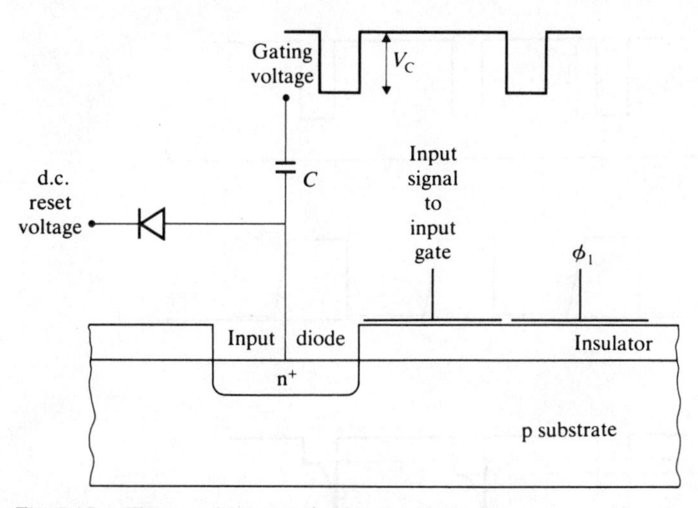

Fig. 5.10 The capacitive metering input technique.

a capacitor for this function. Relative timings of the various waveforms are shown in Fig. 5.11. During the period when ϕ_1 is off, the gating voltage is high so that the input diode voltage is clamped at V_R by the diode connected to the reset voltage. This voltage is higher than the input gate voltage so that the surface potential under the input diode is lower than that under the input gate and the input circuit is non-conducting. Just after ϕ_1 is switched on the input diode potential is pulsed to a low value by the gating voltage so that the surface potential under the input diode is higher than that under the input gate and the input circuit conducts into the drain provided by the potential well under the ϕ_1 electrode. The minimum voltage on the input diode is V_R less the capacitive division of V_c by C and C_s, the series stray capacity around and including the input diode. Conduction continues until the surface potentials under the input diode and input gate become equal when the cut-off condition is reached as described by the threshold voltage V_T between the input gate and input diode. The cut-off point is controlled by the input voltage V_G so that the injected charge is

$$q_s = (C + C_s)(V_G - V_T - V_R) + V_c C \qquad 5.10$$

The charge injected will be linearly proportional to V_G if C_s is constant and V_T is constant or proportional to V_G. In practice non-linearities in C_s associated with the input diode and the reset diode can be designed to cancel. V_T has a dependence on V_G but it is substantially linear. There is also an incomplete charge-transfer non-linearity of the same type as in BB devices (see section 3.7) but the input circuit is only one bucket long so this limit does not appear to be

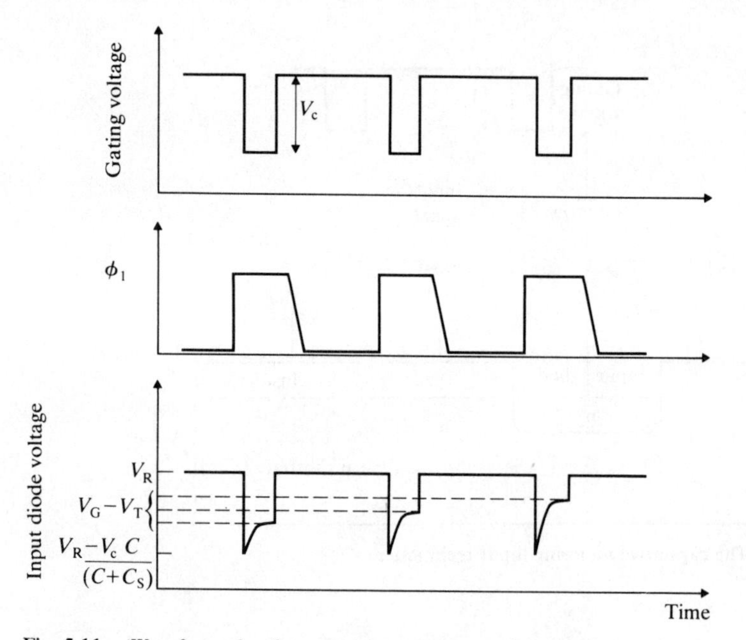

Fig. 5.11 Waveforms for the capacitive metering technique.

significant and is overcome by the fat zero required in the bulk of the CCD. A typical second harmonic distortion is 45 dB below the fundamental with third harmonic distortion 3 dB lower for an input signal which is 50% of full well. Further details have been given by Tozer and Hobson (1976b).

One advantage of this technique is that it enables the working range of the input voltage to be selected by suitable choice of metering capacitor so that it is compatible with the preceding circuitry. A similar modification could be made in fill and spill techniques by altering the area of the receiving electrode and this may well improve their distortion performance.

The capacitive metering technique has some similarities to the threshold insensitive modifications of the fill and spill technique. The charge injected into the CCD is essentially controlled by a change of voltage in both cases, but the former technique self-samples the input voltage by sacrificing the threshold insensitivity. This sacrifice can be overcome by d.c. feedback provided from the output of the CCD to the reset voltage point in the input circuit.

5.3 Output circuits

5.3.1 Destructive sensing with a resistor

Destructive output of the charge from a CCD is performed with a reverse-biased diode as discussed in section 2.16. The simplest output circuit (Fig. 5.12) allows the charge to be converted to a current by flowing through a resistor connected in series with the output-diode bias supply. The current is converted to a voltage

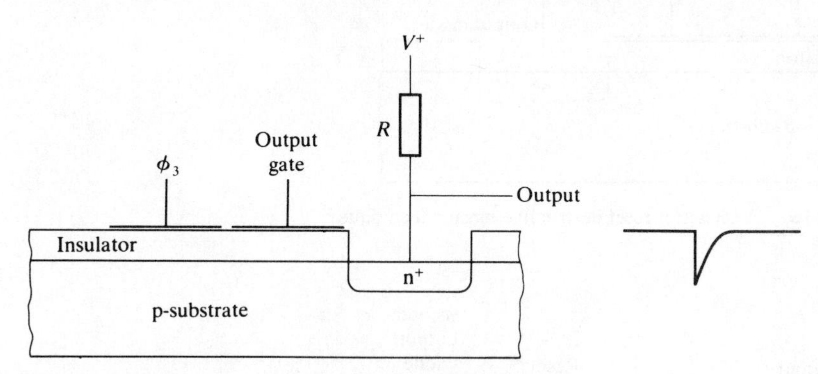

Fig. 5.12 A simple destructive output sensing circuit.

by the resistor. This circuit is unsatisfactory because the output appears as a spike, the details of which are controlled by R, the stray capacity and the rate of charge flow into the output diode from the bulk of the CCD. To be of use, the peak of the spike would have to be 'captured' by a sample and hold circuit and this presents considerable problems in timing accuracy. In addition, unless the discharge time constant is made sufficiently long, the peak is not proportional to the charge output, defined by the area under the spike.

5.3.2 Destructive sensing with charge-reset circuits

The usual destructive way to convert the charge linearly to an output voltage is to collect it on a capacitor and to reset the capacitor voltage in between successive charge outputs as illustrated in Fig. 5.13. The reset switch must be open for the entire time that charge is being transferred into the output diode from ϕ_3 through the region under the output gate. The input impedance of the amplifier following the output must be large enough to avoid any significant discharge of C during the integration period. For a CCD with a maximum charge output of 1 pC the output voltage will be 1 V if $C = 1$ pF. If this charge is collected over a period of 1 μs the input impedance of the following stage must be $\gg 1$ MΩ.

If C is provided 'off-chip' it is difficult to keep its value down to 1 pF and there is a difficulty with spurious pick-up of clock line signals and isolation from output to input. It is preferable to build the reset circuit and the following buffer amplifier 'on-chip' as illustrated in Fig. 5.14. The floating diffusion is simply the reverse-biased output diode but is not externally connected. Instead the reset process is carried out into a second output diode when the floating

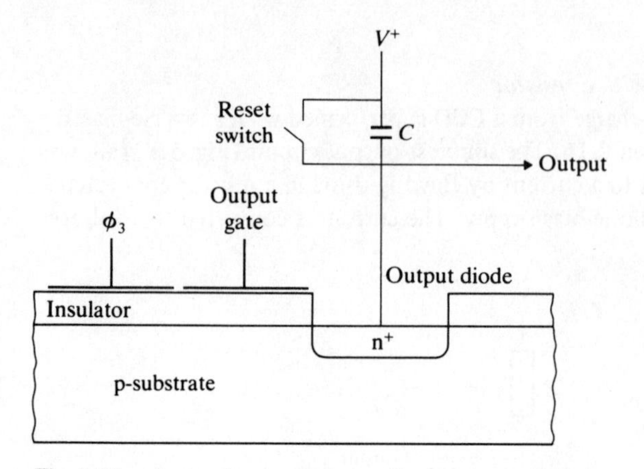

Fig. 5.13 A capacitor reset destructive sensing technique.

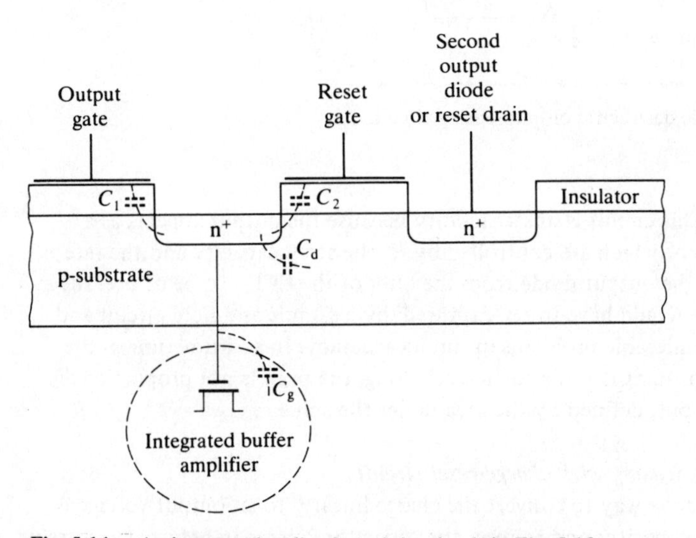

Fig. 5.14 An integrated realization of the circuit in Fig. 5.13.

diffusion surface potential is set to the reset drain potential by the reset gate in between CCD charge outputs. The floating diffusion is connected to the gate of a MOSFET in an integrated buffer amplifier on the same chip. The signal voltage, V_{out}, applied to the input of the buffer amplifier is related to the CCD output charge q_{out} by

$$V_{out} = \frac{q_{out}}{C_1 + C_2 + C_d + C_g}$$

A problem arises with reset signal breakthrough on to the floating diffusion signal unless C_2 can be made sufficiently small. To guard against this possibility an extra d.c.-biased gate is often placed between the floating diffusion and the reset gate. The d.c.-biased output gate performs a similar shielding function against clock signal breakthrough from the last ϕ_3 electrode to the signal on the floating diffusion. An alternative approach is to have a dummy reset gate and a floating diffusion which is not coupled to the CCD channel. The reset breakthrough may then be balanced out if the buffer is a differential amplifier with one input connected to the active output and one input connected to the dummy output.

Care must be taken in the reset timing to avoid any loss of the output charge. The time during which charge leaves the CCD depends on the output gate voltage relative to the clock voltage amplitude as illustrated in Fig. 5.15 (J. Carver,

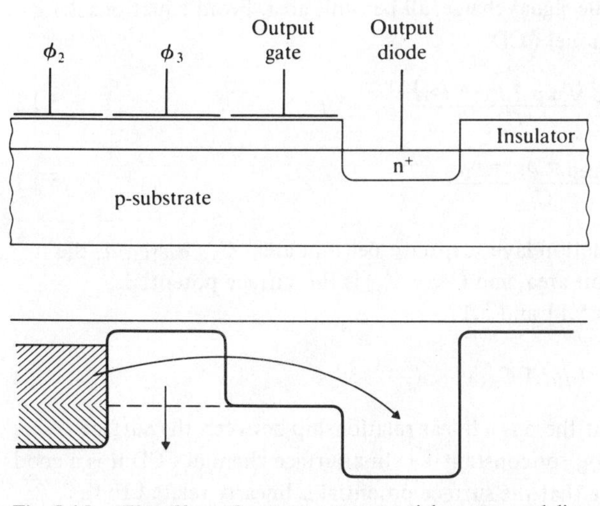

Fig. 5.15 The effect of output gate potential on charge delivery to the output diode.

private communication). The stored charge under ϕ_2 is shown as a large fraction of the full well charge. As the surface potential under ϕ_3 falls, on the rising edge of the ϕ_3 clock waveform, charge will transfer into the output diode potential well until the surface potential under ϕ_3 is equal to the surface potential under the output gate. The remaining charge will not pass beyond the region under ϕ_3 when its surface potential reaches the minimum level and the ϕ_2 surface potential begins to rise. This segment of charge will remain contained by the output gate potential barrier until the ϕ_3 clock voltage starts to fall so that the rising surface potential allows the final transfer into the output diode. Non-linearity would arise for these operating conditions if the reset timing did not allow integration of both segments of charge. Under normal conditions the output gate bias would either be greater than the clock voltage amplitude or much smaller in order to define the charge output timing.

5.3.3 Non-destructive sensing with gate electrodes

Many applications of CCD require several output tapping points so that the destructive sensing techniques cannot be used. All the non-destructive sensing techniques make use of an electrode similar to the clocking electrodes. One suitable type of electrode would be the intermediate phase electrode used for psuedo-two-phase clocking of a three-phase CCD as described in section 2.5.3. The charge stored on either side of the insulator region must be equal in magnitude and opposite in sign so that measurements made on the exposed electrode provide information about the charge underneath the insulator. If the electrode area is A the charge on the electrode is q_B, given by

$$q_B = (\rho_{dep} + \rho_F + \rho_n)A \qquad 5.11$$

ρ_{dep}, ρ_F and ρ_n are respectively the depletion charge $(-N_A el)$, the fixed interface charge and the mobile signal charge, all per unit area. From equations 2.12 or 2.35, for a surface channel CCD,

$$V_B = (V_S - V_o) - \frac{(\rho_{dep} + \rho_F + \rho_n)}{C_{ox}} \qquad 5.12$$

or

$$V_B = \frac{-\rho_{dep}}{C_{dep}} - \frac{(\rho_{dep} + \rho_F + \rho_n)}{C_{ox}} \qquad 5.13$$

C_{dep} is, $2\epsilon_s\epsilon_o/l$, the depletion layer capacity per unit area, C_{ox} is $\epsilon_I\epsilon_o/d$, the insulator capacity per unit area, and $(V_S - V_o)$ is the surface potential.

Combining equations 5.11 and 5.12,

$$V_B = (V_S - V_o) - (q_B/A\ C_{ox}) \qquad 5.14$$

Equation 5.14 shows that there is a linear relationship between the surface potential, $V_S - V_o$, and q_B for constant V_B. In a surface channel CCD it is a good approximation to assume that the surface potential is linearly related to the stored mobile charge for constant bias voltage. To show this equations 5.11 and 5.13 can be combined to give

$$V_B = \frac{(\rho_F + \rho_n)}{C_{dep}} - \frac{q_B}{A}\left(\frac{1}{C_{dep}} + \frac{1}{C_{ox}}\right) \qquad 5.15$$

From equation 5.15, under constant-bias voltage, a change of q_B, Δq_B, is caused by a change of ρ_n, $\Delta\rho_n$, given by

$$\Delta q_B = \frac{A}{[1 + (C_{dep}/C_{ox})]}\Delta\rho_n \qquad 5.16$$

Equation 5.16 shows that there is an equality between $\Delta q_B/A$ and $\Delta\rho_n$ providing $C_{dep} \ll C_{ox}$. This condition is achieved by making the substrate doping as low as possible and the oxide as thin as possible as discussed in sections 2.7 and 2.8. The sensing of Δq_B is usually carried out with a charge amplifier as schematically illustrated in Fig. 5.16. Point X is a virtual earth point defined by the high gain differential amplifier and V_B. Any change of charge on the upper

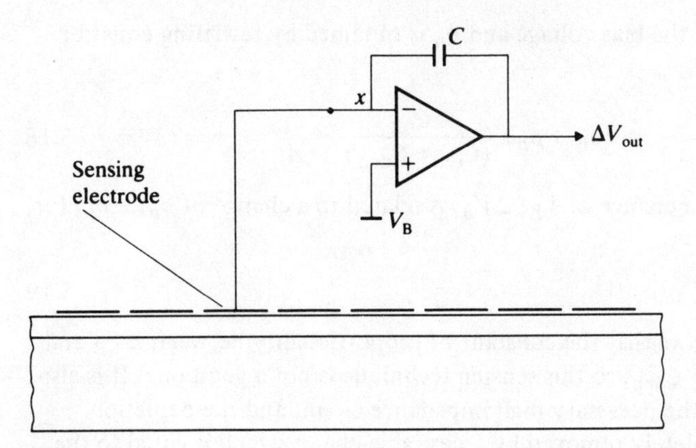

Fig. 5.16 A charge amplifier used non-destructively to sense charge at constant bias voltage.

surface of the sensing electrode is provided by C so that the output voltage is given by

$$\Delta V_{out} = C \Delta q_B \tag{5.17}$$

If the combined residual non-linearities of the non-destructive output sensing, the output charge amplifier and the input system of the CCD are troublesome, they may be further reduced by negative-feedback techniques. For example, if the d.c.-biased charge metering electrode, G_2, of the modified fill and spill input in Fig. 5.4 is used with the same circuit and layout configuration as the output sensors, its non-delayed output can be fed back to the input, as shown in Fig. 5.17, via a differential amplifier so that the fidelity of all the matched output circuits is improved.

Instead of charge sensing at constant voltage it is possible in principle to sense the voltage change, as mobile charge passes underneath an isolated sensing electrode. In this case q_B remains constant in the sensing operation. The

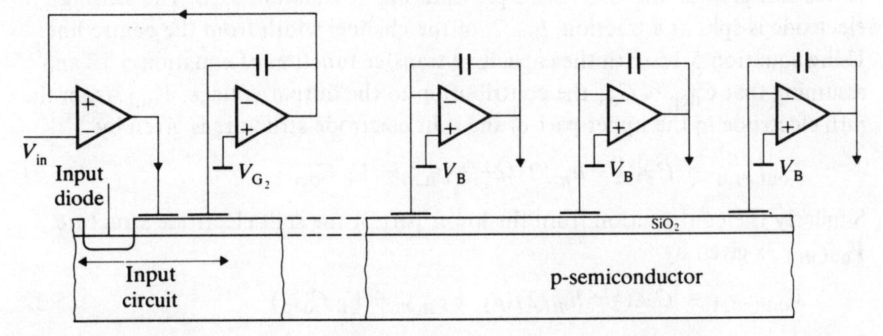

Fig. 5.17 Feedback correction for constant-voltage charge sensing circuits.

relationship between the bias voltage and ρ_n is obtained by rewriting equation 5.15 as

$$V_B \frac{C_{dep}\, C_{ox}}{(C_{o,x} + C_{dep})} = (\rho_F + \rho_n) \frac{C_{ox}}{(C_{ox} + C_{dep})} - \frac{q_B}{A} \quad 5.18$$

From equation 5.18 a change of V_B, ΔV_B, is related to a change of ρ_n, $\Delta\rho_n$, for q_B constant by

$$\Delta V_B = \Delta\rho_n / C_{dep} \quad 5.19$$

Equation 5.19 shows that the constant of proportionality between ΔV_B and $\Delta\rho_n$ is the non-linear C_{dep}, so this sensing technique is not a good one. It is also difficult to produce the necessary high impedance circuit and the depletion layer would be completely removed by a desirable charge which is equal to the depleted charge of the empty well. Therefore this sensing technique would become inoperative for stored charge much less than the saturation value (see equation 2.26), and would allow recombination of the charge with majority carriers.

5.3.4 Split-electrode tapping

Later considerations of matched filtering will require the generation of a signal which is the sum of signals sensed at many sequential points of the CCD. Before summation each sensed signal will have to be multiplied by a constant which is, in general, different at each sensing point. This process is referred to as convolution. The output signal, V_{out}, is given by

$$V_{out} \propto \sum_m h_m\, \rho_{n,m} \quad 5.20$$

h_m is referred to as the tap weight and can take any value between 0 and 1. $\rho_{n,m}$ is the signal charge per unit area at the mth sensing node. Providing h_m can be fixed at the production stage of the CCD the process of equation 5.20 can be simply and cheaply carried out by dividing one set of electrodes into two parts as shown in Fig. 5.18. The difference of the signals from the upper and lower halves will provide the analogue representation of equation 5.20. The sensing electrode is split at a fraction, $h_m/2$, of the channel width from the centre line. Using equation 5.15 with the capacitive transfer function of equation 5.17 and assuming that $C_{dep} \ll C_{ox}$ the contribution to the output voltage, V_{out}, from the mth electrode in the upper part of the split electrode structure is given by

$$V_{out,m,u} = CA(\tfrac{1}{2} - h_m/2)(\rho_F + \rho_{n,m} - V_B\, C_{dep}) \quad 5.21$$

Similarly the contribution from the lower part of the spilt electrode structure, $V_{out\ m,l}$, is given by

$$V_{out,\,m,\,l} = CA(\tfrac{1}{2} + h_m/2)(\rho_F + \rho_{n,\,m} - V_B\, C_{dep}) \quad 5.22$$

The output signal, V_{out}, is given by

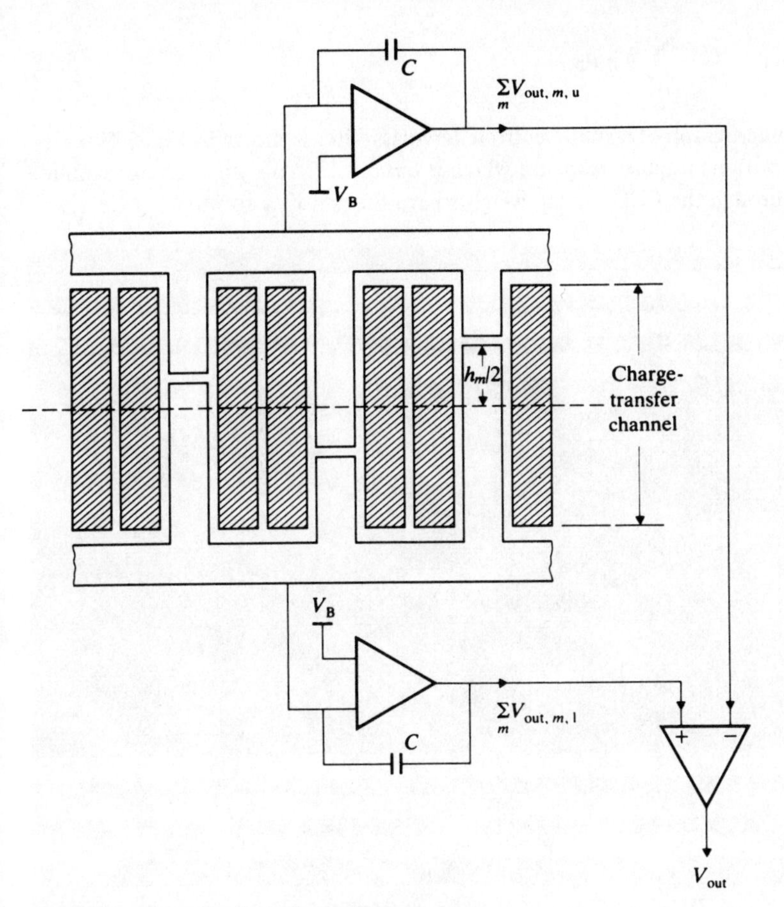

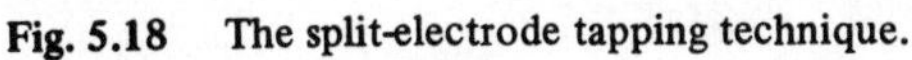

Fig. 5.18 The split-electrode tapping technique.

$$V_{\text{out}} = \sum_m V_{\text{out}, m, \text{l}} - \sum_m V_{\text{out}, m, \text{u}} \qquad 5.23$$

i.e.
$$V_{\text{out}} = CA \sum_m h_m \rho_{\text{n}, m} + CA(\rho_{\text{F}} - V_{\text{B}}\, C_{\text{dep}}) \sum_m h_m \qquad 5.24$$

The second term on the right-hand side of equation 5.24 is a constant independent of the signal level. It is zero if $\sum_m h_m = 0$. If this condition is fulfilled, the independence of V_{out} on V_{B} in equation 5.24 shows that the split electrode structure may be biased with the normal clock waveform because the clock signal is balanced out by the differential sensing process. In this case equation 5.24 simplifies to the desired form in equation 5.20.

i.e. $$V_{out} = CA \sum_{m} h_m \rho_{n,m} \qquad 5.25$$

A photomicrograph of a split-electrode low-pass filter is shown in Fig. 5.19 together with its impulse response which is obtained from a single charge sample passing through the CCD. Its tap weights have the $(\sin x)/x$ form.

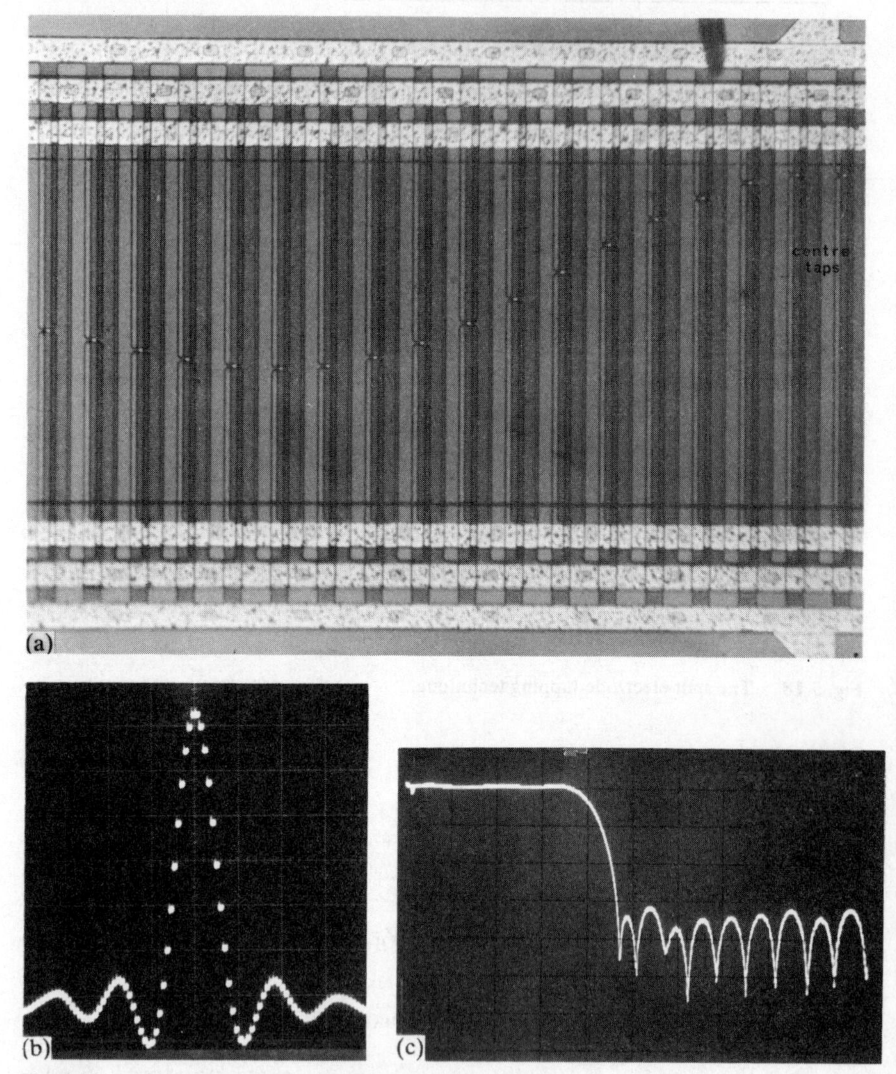

Fig. 5.19 (a) A photomicrograph of part of a split-electrode low-pass filter and (b) its impulse response. The lateral position of each electrode split may be directly compared with the impulse response on a point by point basis. The peak of the impulse response corresponds to the centre taps of the photomicrograph. The frequency response is shown in (c) for completeness and will be referred to later in section 6.2. Its vertical scale is 10 dB/major division and the horizontal scale is 1 kHz/major division. (Reproduced by permission of GEC, Hirst Research Centre).

5.3.5 The distributed floating-gate amplifier

In this technique, the change of surface potential, caused by signal charge under constant bias-voltage conditions, is used to control the flow of current through a MOS field effect transistor. A schematic cross-section of the construction is given in Fig. 5.20, where the direction of charge transfer is into the page. The signal charge modifies the surface potential which, by capacitive division, modifies the

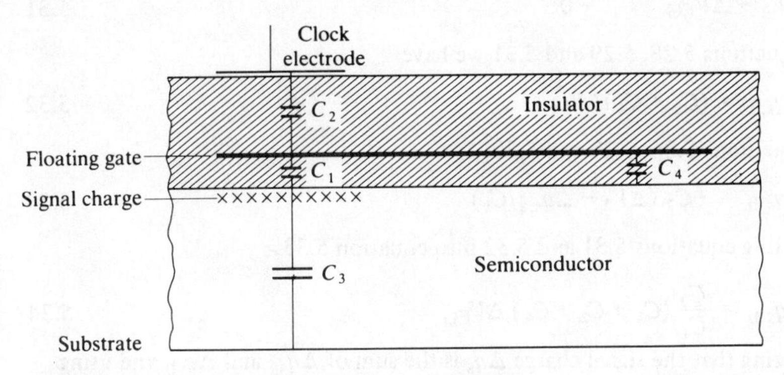

Fig. 5.20 A cross-section of the floating-gate amplifier.

floating gate potential. The extension of the floating-gate outside the CCD region allows it also to act as the gate of a MOSFET whose source and drain contacts can be fabricated with the same technology but are not shown here. Also shown on Fig. 5.20 are the capacities associated with each region of the floating-gate device. When a signal charge, $\Delta q_S (= \Delta q_{s,a} + \Delta q_{s,b})$, is present it causes a redistribution of charge, Δq_G, into the FET region without changing the net charge on the floating gate. As illustrated in Fig. 5.21 there is also a

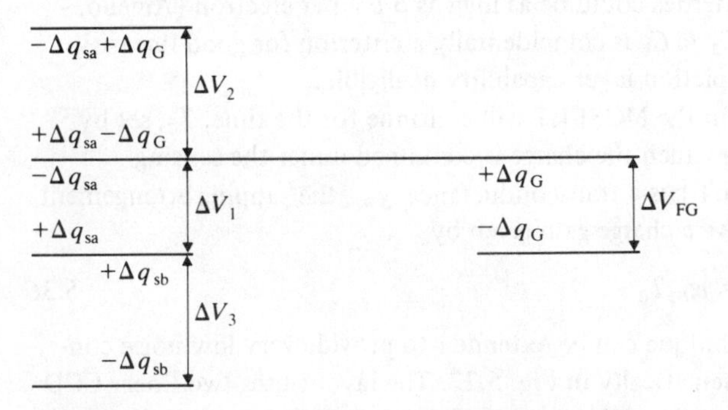

Fig. 5.21 Charge conditions in a floating-gate amplifier.

charge flow to the clock electrode, the substrate and the interface in the FET region which are maintained at constant voltage. The changes of inter-electrode voltage must satisfy the following relationships:

$$\Delta q_{s,b} = C_3\,\Delta V_3 \qquad 5.26$$

$$-\Delta q_{s,a} = C_1\,\Delta V_1 \qquad 5.27$$

$$-\Delta q_{s,a} + \Delta q_G = C_2\,\Delta V_2 \qquad 5.28$$

$$\Delta q_G = C_4\,\Delta V_{FG} \qquad 5.29$$

$$\Delta V_1 + \Delta V_2 + \Delta V_3 = 0 \qquad 5.30$$

$$\Delta V_2 + \Delta V_{FG} = 0 \qquad 5.31$$

From equations 5.28, 5.29 and 5.31 we have

$$\Delta q_{s,a} = (C_2 + C_4)\Delta V_{FG} \qquad 5.32$$

From equations 5.26, 5.27 and 5.30,

$$\Delta q_{s,b} = -C_3\,(\Delta V_2 - \Delta q_{s,a}/C_1) \qquad 5.33$$

Substituting equations 5.31 and 5.32 into equation 5.33,

$$\Delta q_{s,b} = \frac{C_3}{C_1}(C_1 + C_2 + C_4)\,\Delta V_{FG} \qquad 5.34$$

Recognizing that the signal charge Δq_s is the sum of Δq_{sa} and Δq_{sb} and using equations 5.32 and 5.34, the transfer function is

$$\frac{\Delta V_{FG}}{\Delta q_s} = \frac{1}{\left[(C_2 + C_4) + \frac{C_3}{C_1}(C_1 + C_2 + C_4)\right]} \qquad 5.35$$

For high sensitivity C_3 should be small (weak doping and wide depletion layer), C_1 should be large (thin oxide), and C_2 and C_4 should be as small as possible consistent with other design and technological criteria. If $(C_2 + C_4)$ is 1 pf for example, $\frac{\Delta V_{FG}}{\Delta q_s}$ is 10^{12} V C^{-1} which is equivalent to 0.16 μV per electron. It has been claimed that sensitivities could be as high as 5 μV per electron (Amelio, 1974). The inequality $C_3 \ll C_1$ is coincidentally a criterion for good linearity because it makes the depletion layer capability negligible.

The current flow from the MOSFET will continue for the time, T_c, set by the clock period, during which the charge is contained under the sensing electrode. If the MOSFET has a transconductance, g_m, the tapping arrangement can be considered to have a charge gain given by

$$A_Q = (\Delta V_{FG}/\Delta g_s)g_m T_c \qquad 5.36$$

The floating gate technique can be extended to provide very low noise conversion, as indicated schematically in Fig. 5.22. The layout uses two linear CCD in parallel. Each floating gate of the input shift register non-destructively detects its own charge and injects charge into the corresponding well of the output shift register. After transfer to the next floating gate, the same charge packet in the input register causes injection of charge into the corresponding well of the output shift register which also contains the transferred charge from the first injection. The process repeats at each floating gate and synchronism is maintained between the successively larger charge packets in the output register and their corresponding charge packets in the input shift register. Noise is also generated by the charge injections to the second output register but successive inputs only add in a root mean square or random walk fashion in contrast with the coherent addition of the desirable charge inputs. If there are p injections from the input to output registers the signal will be p times larger at the output whereas the noise will only be $\sqrt{p}$ times larger so that the improvement in signal/noise power ratio compared with that for one output gate is p.

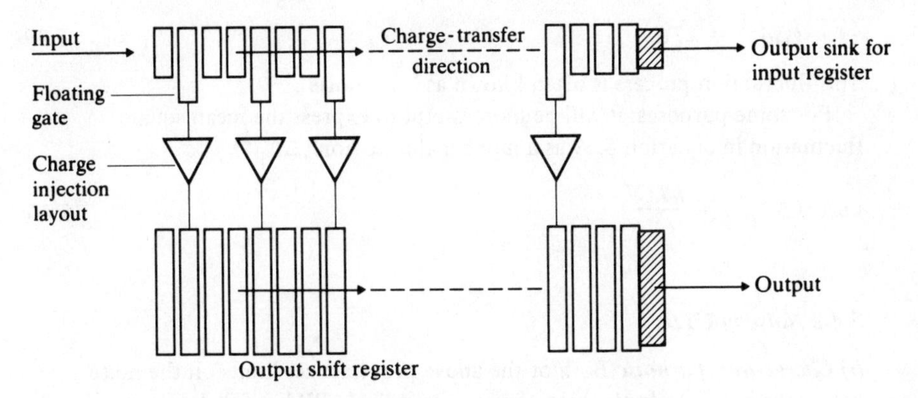

Fig. 5.22 A distributed floating-gate amplifier (DFGA).

5.4 Noise considerations

5.4.1 Relevant noise sources

Two sources of electrical fluctuation (Bell, 1960) need introduction before CTD noise can be discussed.

(i) Shot noise Electrons or holes in a semiconductor exhibit random current fluctuations called shot noise. The source of the fluctuations is the random thermal velocity which is superimposed on the mean flow velocity so that the passage of charge carriers, through a plane which is transverse to the current flow, occurs at random instants of time. If the current corresponds with N particles the mean square fluctuation is given by

$$\langle \Delta N^2 \rangle = N \qquad 5.37$$

Shot noise components combine randomly. That is, the total mean square noise is the sum of the mean square noise in each component.

(ii) Johnson noise The fluctuating energies of the charge carriers cause another component of electrical noise, called Johnson noise, which is relevant to the storage of charge in a capacitor. The fluctuation in the stored energy of the capacitor is given by

$$\tfrac{1}{2}kT = \tfrac{1}{2}C\langle \Delta V^2 \rangle$$

The fluctuations cause a voltage fluctuation at the capacitor terminals which is given by

$$\langle \Delta V^2 \rangle = kT/C. \qquad 5.38$$

When the capacitor is connected to other circuitry so that charge flow can occur, there is a charge fluctuation given by

$$\langle \Delta q^2 \rangle = kTC. \qquad 5.39$$

This fluctuation process is often known as 'kTC noise'.

For some purposes, it will be more useful to express the mean square fluctuation in equation 5.39 as a number fluctuation $\langle \Delta N_p{}^2 \rangle$.

$$\text{i.e.} \quad \langle \Delta N_p{}^2 \rangle = \frac{kTC}{e^2} \qquad 5.40$$

5.4.2 Noise in CTDs

(i) Charge-transfer noise Both of the above processes set limits on the noise behaviour in CTDs. In the case of charge transfer, a BB has full Johnson noise at each transfer because the transfer terminates according to the voltage on the emptying capacitor. In a device with n transfers, the total mean square number fluctuation will be

$$\langle \Delta N_{tot}{}^2 \rangle = \frac{n\,kTC_{BB}}{e^2} \qquad 5.41$$

where C_{BB} is the BB storage capacitor. The CCD scores a great advantage over the BB in this respect particularly for low light level imaging applications. Even though the charge stored in its capacitive potential wells will exhibit Johnson noise it has no control over the transfer process owing to the way charge is 'poured' from well to well. There will be a residual shot noise in the incomplete charge-transfer fraction, ϵ. If the signal charge number is N_S per well, on each transfer there will be a mean square fluctuation of ϵN_S associated with the received charge smearing from the preceding packet and a further ϵN_S associated with lost charge smearing to the following packet. The charge number fluctuation after n transfers is

$$\langle \Delta N_\epsilon{}^2 \rangle = 2n\,\epsilon\,N_S \qquad 5.42$$

A third contribution to the fluctuations in both BB and CCD arises from surface states and spurious energy levels. As discussed in section 4.2, the release rate of minority carriers depends on the position of the spurious energy levels in the forbidden energy gap of the semiconductor. Those states whose filling and emptying times are either much faster or much slower than the charge-transfer period allowed by the clock frequency do not contribute to any fluctuations. It is only states whose time constants are comparable with the transfer period that give rise to a number fluctuation in any charge packet. The dependence of the time constant of equation 4.4 on the thermal energy intuitively indicates that the energy range of importance will be approximately kT wide and centred on the energy where τ_{empty} is equal to the charge-transfer time. Assuming complete filling of the surface states of density N_{SS} m^{-2} eV^{-1}, as appropriate for operation with a fat zero, the charge number fluctuation will be

$$\langle \Delta N_{SS}{}^2 \rangle = N_{SS} \frac{kT}{e} A \tag{5.43}$$

where A is the area of the device. e in the denominator of equation 5.43 arises from the practical units used for N_{SS}. Similar fluctuations arise in buried channel devices but the spurious state density is much smaller and it is usually associated with discrete energy levels rather than with a continuum. This causes peaks in the output noise level as the clock frequency is changed (Mohsen and Tompsett, 1974).

The effect of transfer noise on a signal depends on the signal frequency. At low frequencies successive signal charges are almost equal and the successive random noise fluctuations tend to cancel owing to the equality of the charge smearing loss from one charge packet and the charge smearing gain in the next packet. At high frequencies, as the Nyquist limit is approached, successive signal charges may be quite different so that this cancellation feature is lost. A correct treatment of this process by Fourier transformation of the time domain response and low-pass filtering to exclude frequencies above the Nyquist limit gives a noise spectrum described by Thornber and Tompsett (1973):

$$\langle N(\mathrm{f})^2 \rangle = \langle N(\mathrm{o})^2 \rangle \left(1 - \frac{\cos \pi f}{f_{\mathrm{Nyquist}}}\right) \tag{5.44}$$

$\langle \mathrm{N(o)}^2 \rangle$ is the number fluctuation that would occur if there was no correlation between any of the successive noise fluctuations so that all their mean square values added. $\langle N(f)^2 \rangle$ is the number fluctuation for an input frequency f, and f_{Nyquist} is the Nyquist limiting frequency.

(ii) Input noise Noise in the input circuit depends on the charge injection process. For the FET input, the input current will exhibit shot noise so that the fluctuation in the signal charge is given by

$$\langle \Delta N_{\mathrm{FET}}{}^2 \rangle = N_{\mathrm{S}} \tag{5.45}$$

A similar result will be obtained for optical input owing to the shot noise of the photons.

In the diode cut-off mode and the various fill and spill modes the charge injected is stored in a capacitive potential well at a level set by the corresponding voltage. The noise will be given by

$$\langle \Delta N^2{}_{\text{diode cut-off}} \rangle = \langle \Delta N^2{}_{\text{fill \& spill}} \rangle = \frac{kTC_{\mathrm{cc}}}{e^2} \tag{5.46}$$

where C_{cc} is the capacity of the initial charge containment well. In the case of the diode cut-off there will be an excess partition noise as mentioned in section 5.2.

(iii) Output noise Similar comments apply at the output where charge is collected in a capacity, C_{reset}. In this case the reset charge number fluctuation is given by

$$\langle \Delta N^2{}_{reset} \rangle = \frac{kTC_{reset}}{e^2} \tag{5.47}$$

$\langle \Delta N^2{}_{reset} \rangle^{\frac{1}{2}}$ causes a corresponding voltage uncertainty on the gate of the output buffer amplifier.

(iv) Numerical estimates For numerical estimates, a full well might contain 10^7 minority carriers with an area of 10^{-9} m^{-2}. Both of these quantities may be an order of magnitude higher than for some CCDs. Comparisons will be made with the number of transfer events, $n = 200$, the incomplete charge-transfer fraction, $n\epsilon = 0.1$ and the surface $N_{SS} = 10^{14}$ m^{-2} eV^{-1}. Capacities will only include the oxide with a thickness of 0.1 μm.
CCD charge-transfer noise From equation 5.42,

$$\langle \Delta N_\epsilon^2 \rangle^{\frac{1}{2}} = (2n\epsilon N_S)^{\frac{1}{2}} = 1400 \text{ electrons} \tag{5.48}$$

which is 1.4 x 10^{-4} of the saturation charge. For a much smaller signal charge of 10^3 electrons,

$$\langle \Delta N_\epsilon^2 \rangle^{\frac{1}{2}} = 14 \text{ electrons} \tag{5.49}$$

which illustrates the low noise property of CCD devices.
Surface state noise From equation 5.43,

$$\langle \Delta N_{SS}{}^2 \rangle^{\frac{1}{2}} = \left(N_{SS} \frac{kT}{e} A \right)^{\frac{1}{2}} = 50 \text{ electrons/transfer}$$

$$= 700 \text{ electrons per 200 transfers} \tag{5.50}$$

which is 0.7 x 10^{-4} of the saturation charge.

In a buried channel CCD with 2 x 10^{17} m^{-3} traps the number of states per well is ~400 so the corresponding fluctuation is

$$\langle \Delta N_{SS}{}^2 \rangle^{\frac{1}{2}}{}_{\text{buried channel}} = 20 \text{ electrons per transfer}$$

$$= 280 \text{ electrons per 200 transfers} \tag{5.51}$$

which is only 2.8 x 10^{-5} of the saturation charge.
Input and output noise From equation 5.20 the FET input noise for a full well is

$$\langle \Delta N_{FET}^2 \rangle^{\frac{1}{2}} = (N_S)^{\frac{1}{2}} \simeq 3000 \text{ electrons} \tag{5.52}$$

The other CCD input and output mechanisms limited by Johnson noise as shown in equations 5.46 and 5.47 have an r.m.s. fluctuation given by

$$\langle \Delta N^2 \rangle^{\frac{1}{2}}_{\text{input, output}} = (kTC_{\text{input, output}})^{\frac{1}{2}}/e = 185 \text{ electrons} \qquad 5.53$$

Equation 5.53 shows the much better noise performance of Johnson noise limited interface techniques (~2×10^{-5} of saturation charge) compared with shot noise limited techniques such as the FET input (3×10^{-4} of saturation charge). Equations 5.49 and 5.51 indicate the low noise capability of CCDs at low signal levels such as those occurring in low light level application of CCD television cameras. The output technique is a major limitation for such applications. The floating gate technique described in section 5.3.5 does not receive or lose charge so it is essentially capable of nearly noiseless operation owing to the absence of Johnson noise and shot noise. Residual noise in the buffer amplifier can be overcome to some extent by use of the distributed floating-gate amplifier.

5.4.3 Correlated double sampling

An alternative way of reducing the output circuit noise is the 'correlated double sampling' (CDS) technique. Even though the charge uncertainty introduced by the reset process causes a corresponding voltage error the subsequent Johnson noise fluctuations are not instantaneous. If the output voltage is measured just before and just after the signal charge enters the output diffusion (Fig. 5.23) the difference voltage will be proportional to the signal charge plus a much smaller fluctuation. Further implementation details have been given by White *et al* (1974) and Brodersen and Emmons (1976). An additional advantage of this technique is the subtraction of clock pick-up signals from the output.

The above descriptions of noise processes in CCDs have been based on simplified concepts and have neglected other noise sources. Further details and experimental measurements which substantially confirm the above estimates

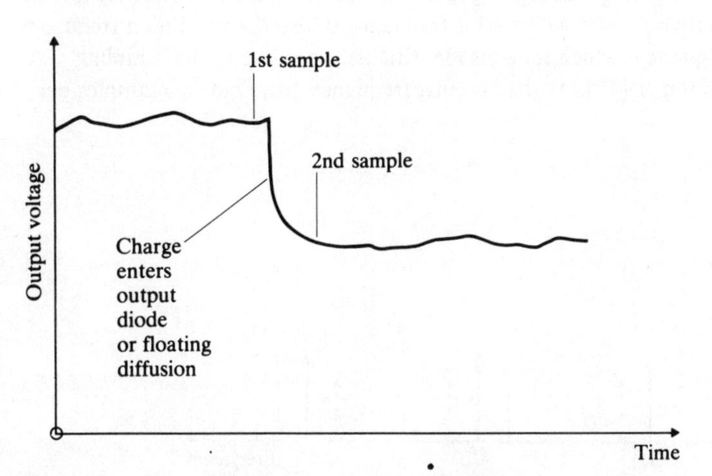

Fig. 5.23 The sampling times for correlated double sampling.

have been given by Carnes and Kosonocky (1972), Carnes *et al* (1973) and Thornber (1974). Even though the maximum charge per well is only $\sim 10^7$ electrons the CCD is capable of dynamic ranges in excess of 80 dB. BB devices appear to be limited to a dynamic range of less than 70 dB.

5.5 Sampled-data errors: aliasing and the Nyquist frequency

Sampling of the input signal eliminates information in a way which allows a variety of different signals to be stored as the same signal in the CTD. Fig. 5.24

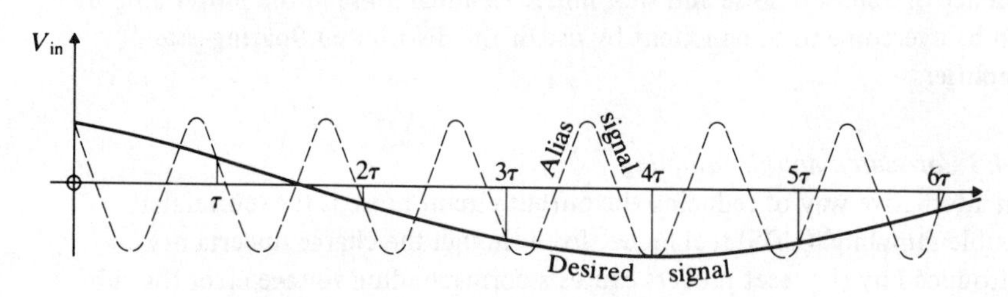

Fig. 5.24 The aliasing process at the sampling input to a CTD.

illustrates this behaviour for two signals with different frequencies which are undergoing delta-function sampling at intervals τ. The corresponding representation in the frequency domain is shown in Fig. 5.25 for a desirable signal of frequency, f_s, and a sampling frequency, $f_c(= 1/\tau)$. The sampling behaviour is the same as the frequency conversion process of a superhet receiver operating with a local oscillator which is rich in harmonics. All signals with frequencies given by $nf_c \pm f_s$, where n is an integer, will cause a signal to be stored in the CTD with frequency f_s. This process is known as aliasing. Clearly, the CTD cannot be used for unambiguous signal processing unless the input frequency is limited to less than $f_c/2$ $(= 1/2\tau)$. This frequency is also the maximum frequency, or Nyquist frequency, which remains identifiable according to the sampling theorem (Shannon, 1949). At the Nyquist frequency there are two samples per cycle.

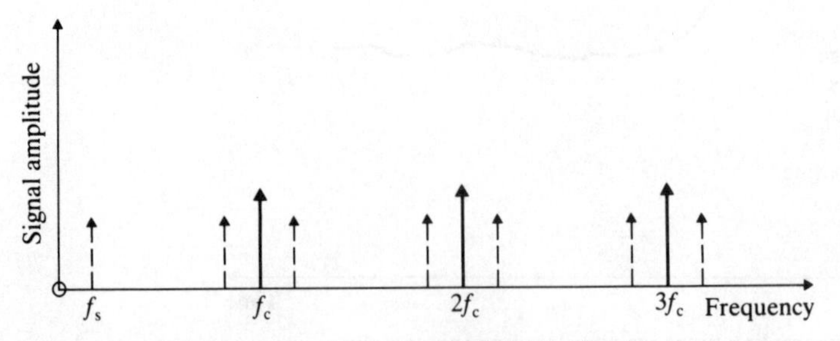

Fig. 5.25 The spectral content of a sampled data signal.

Even if the signal bandwidth is less than $f_c/2$, wide band noise will be down-converted and degrade the signal/noise performance. Some form of low-pass filtering is required to overcome these difficulties, but for practical reasons, it should be small compared with the CTD and be integrated on the same chip. One simple implementation arises from the fill and spill input technique. The signal sampling occupies the finite time allowed for the spill process. Signals whose frequency is greater than the Nyquist frequency are essentially peak detected if the sampling window is made a large fraction of τ. If the process was a perfect peak detection the aliasing would be replaced by a simple d.c. offset. However, the equilibration time of charge in the spill process is finite and the details are more complicated, but overall, a reduction of aliasing components in excess of 10 dB appears possible by correct design of the sampling window. Greater detail has been provided by Emmons *et al* (1975).

An alternative approach is to design the input circuit as a transversal, low-pass, pre-filter. The input charge is transferred into the CTD along several parallel channels. Some practical layouts have been described by Berger and Coutures (1976). Each channel receives its input sample during a different staggered window within one sampling period. Each path suffers a corresponding time delay so that they are all combined coincidentally as they enter the main CTD transfer channel. The input channel widths are proportional to the weightings required to give the appropriate low-pass characteristic so that low frequency desirable signals add in phase, but aliasing signals have out-of-phase components and cancel. In addition to the greater clock driving complexity, this technique does not completely remove the aliasing problem; it simply moves it to a higher frequency.

When the aliasing problem has been overcome, there is a residual distortion problem caused by the finite sampling time and the consequent uncertainty in the effective time of sampling, as the desired signal level changes. This is influenced by the finite time of charge transfer through the input and becomes more serious at frequencies approaching the Nyquist limit. On rising signals the tendency is to sample at a late time of the sampling window, while on falling signals the sampling occurs relatively early, or *vice-versa*. Without care several per cent harmonic distortion may readily occur (Sequin and Mohsen, 1975).

5.6 Signal reconstruction and sampling attenuation

The output from a CTD is a pulsed unidirectional 'carrier' at the clock frequency, which is amplitude modulated by the desirable signal. Usually it has a constant level in between reset operation so that it is rather like an amplitude modulated square wave which has been half-wave rectified as shown in Fig. 5.26. The spectral content of the signal has the same form as illustrated in Fig. 5.25. In the simplest reconstruction scheme low-pass filtering is used with a cut-off frequency typically at half the clock frequency. Such a technique never completely cancels the unwanted clock frequencies and sidebands. A better technique in principle is to split the input signal into two antiphase components and then pass them in parallel through two identical CTDs. One further inversion after one of the CTD outputs followed by addition of the two signals will cause the unwanted clock frequencies to cancel while the desirable signal frequencies add. The limitation on clock signal rejection is only set by the closeness of matching between the two CTDs (Butler *et al*, 1973). The technique aldo removes unwanted d.c. offset signals but is expensive in use of semiconductor chip area.

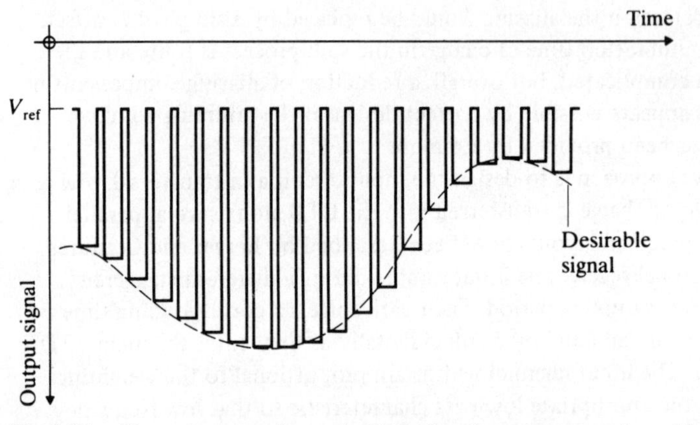

Fig. 5.26 The output waveform of a CTD after the reset or sample and hold process but

Even if perfect low-pass filtering was carried out there would still be signal degradation in the output process for any finite width of the individual square pulses in Fig. 5.26. The sampled data signal processing introduces a frequency response which can be calculated simply from the impulse response of the output circuit shown in Fig. 5.27. One output sample of unit amplitude is considered at a time $N\tau$ after entry to the CTD. τ is the time between samples. The exit of the charge sample from the CTD provides the impulse for the next circuit. It holds the output signal constant for a time ΔT and this is the impulse response of the output circuit. The frequency response is the Fourier transform of the impulse response and is given by

$$f(\omega) = \Delta T \exp(-j\omega N\tau) \exp\left(-\frac{j\omega\Delta T}{2}\right) \left[\frac{\sin(\omega\Delta T/2)}{(\omega\Delta T/2)}\right] \qquad 5.54$$

The term $\exp(-j\omega N\tau)$ is the phase shift caused by the delay of the CTD. The term $\exp(-j\omega\Delta T/2)$ is an extra delay associated with the signal reconstruction process and is caused by the shift of the centre of the output pulse, in Fig. 5.27, relative to the impulse causing it. Equation 5.54 also shows that the frequency response shows an amplitude modification of the $(\sin x)/x$ type.

If the impulse in Fig. 5.27 had an amplitude of $V_0 \sin(m\omega\tau)$, where $m\tau$ is the time variable, the contribution to the output signal from one sample would have been

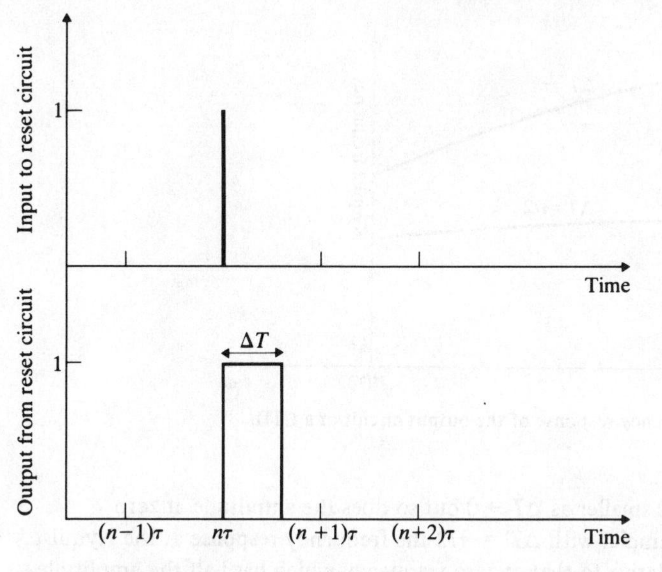

Fig. 5.27 The impulse response of the output circuit of a CTD.

$$(V_{\text{out}})_{\text{sampling}} = V_0\,\Delta T \sin(m\omega\tau)\exp(-\mathrm{j}\omega N\tau)\exp\left(-\frac{\mathrm{j}\omega\Delta T}{2}\right)\left[\frac{\sin(\omega\Delta T/2)}{(\omega\Delta T/2)}\right] \qquad 5.55$$

If the delay had been carried out by a continuous process and followed by low-pass filtering as in the present case the signal contribution to this implied integration process would have been

$$(V_{\text{out}})_{\text{cont}} = V_0\,\tau \sin(m\omega\tau)\exp(-\mathrm{j}\omega N\tau) \qquad 5.56$$

Even though the response in equation 5.56 is a continuous one, the time variable has been written in the same form as in equation 5.55 to facilitate the following comparison. The ratio of $[V_{\text{out}}]_{\text{sampling}}$ to $(V_{\text{out}})_{\text{cont}}$ is the effective frequency response $[f(\omega)]_{\text{sampling}}$, of the sampled data output circuit.

i.e. $$[f(\omega)]_{\text{sampling}} = \frac{\Delta T}{\tau}\exp\left(-\frac{\mathrm{j}\omega\Delta T}{2}\right)\left[\frac{\sin(\omega\Delta T/2)}{(\omega\Delta T/2)}\right] \qquad 5.57$$

Equation 5.57 shows that the output has the familiar $\frac{(\sin x)}{x}$ dependence on frequency caused simply by the sampling process. The signal amplitude is a maximum when ΔT has its largest value which is equal to τ, but the frequency response at the Nyquist frequency, where $\omega = \pi/\tau$, is 0.637 of its value for zero frequency as shown in Fig. 5.28. The dependence of this error on

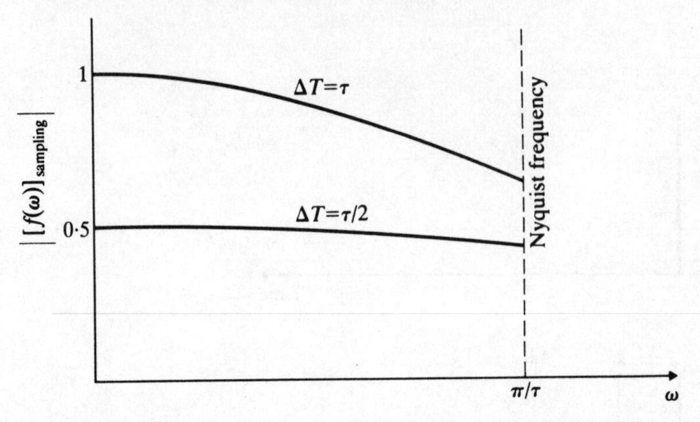

Fig. 5.28 The frequency response of the output circuit of a CTD.

frequency becomes smaller as $\Delta T \to 0$ but so does the amplitude at zero frequency. For example, with $\Delta T = \tau/2$ the frequency response at the Nyquist frequency is 0.9 relative to that at zero frequency which has half the amplitude compared with the previous case. There is a corresponding reduction in the extra delay, $\Delta T/2$, implied by the term $\exp(-j\omega\Delta T/2)$ as $\Delta T \to 0$. This term is not necessarily a defect because it only causes a frequency independent delay.

6

Delay Line Signal Processing

6.1 Introduction

CTDs find several applications as simple delay lines for times of less than one second. They can be used for the delay circuits of PAL colour television systems and, with appropriate control of the clock frequency, they can be used to correct speed errors in audio and video tape-recorders. The electronically controllable delay feature is also exploited for beam steering and focusing of ultrasonic and sonar signals using phased array systems. The transmitting or receiving interface consists of an array of parallel transducers. Each one is fed by a CTD delay line whose delay linearly increases across the array (Fig. 6.1). Beam steering is accomplished by varying the common clock rate of all the CTD delay lines. Focusing is achieved in a similar manner, but in this case the delay times have a quadratic variation which is symmetrical about the central channel. Further details of this system have been given by Melen *et al* (1975) and White and Lampe (1975).

Time axis conversion may be carried out for blocks of data no greater than the storage capacity of the CCD by using different clock frequencies for the entry and exit of data. For example, a transient recorder may take advantage of the high clock frequencies, possibly with buried channel devices, for data collection followed by a slow read-out to the following circuitry.

Other forms of signal processing with CTDs essentially involve combining signals with a regular spacing in time. In the following sections the possibilities offered in the time and frequency domain will be outlined in simple terms. Initially the discussion will neglect the sampled data nature of the various signals. In the frequency domain descriptions it is simply necessary to recognize that a delay of time τ causes a signal of frequency ω to have a phase shift of $\omega\tau$ so that the transfer function is $\exp(-j\omega\tau)$. For example, a delay line with attenuation α, and input signal $V_{in} \cos \omega t$, delivers an output signal given by

$$\begin{aligned} V_{out} &= \mathrm{Re}[\alpha V_{in} \exp(-j\omega\tau) \exp(j\omega t)] \\ &= \alpha V_{in} \cos \omega(t-\tau) \end{aligned} \qquad 6.1$$

6.2 The frequency response of tranversal filters

The simplest example of a transversal filter is the low frequency cancelling filter illustrated in Fig. 6.2(a) which may be used to remove low frequencies in moving target indicator (MTI) radar systems. It has transmission zeros when $\omega = 2m\pi/\tau$ where m is an integer (Fig. 6.2(b)). The phase response is linear with ω and has a $\pi/2$ phase lag at zero frequency. The transfer function is

$$V_{out}/V_{in} = \exp(-j\omega\tau) - 1$$
$$= 2\sin(\omega\tau/2)\exp[-j(\omega\tau/2 + \pi/2)] \qquad 6.2$$

As $\omega \to 0$, $V_{out}/V_{in} \to -j\omega\tau$ which is analogous to a lumped-circuit differentiator.

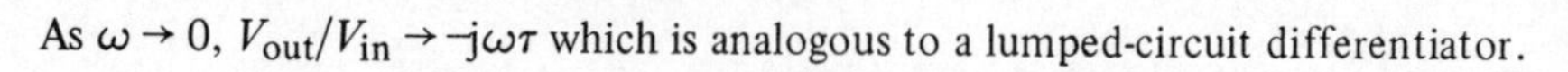

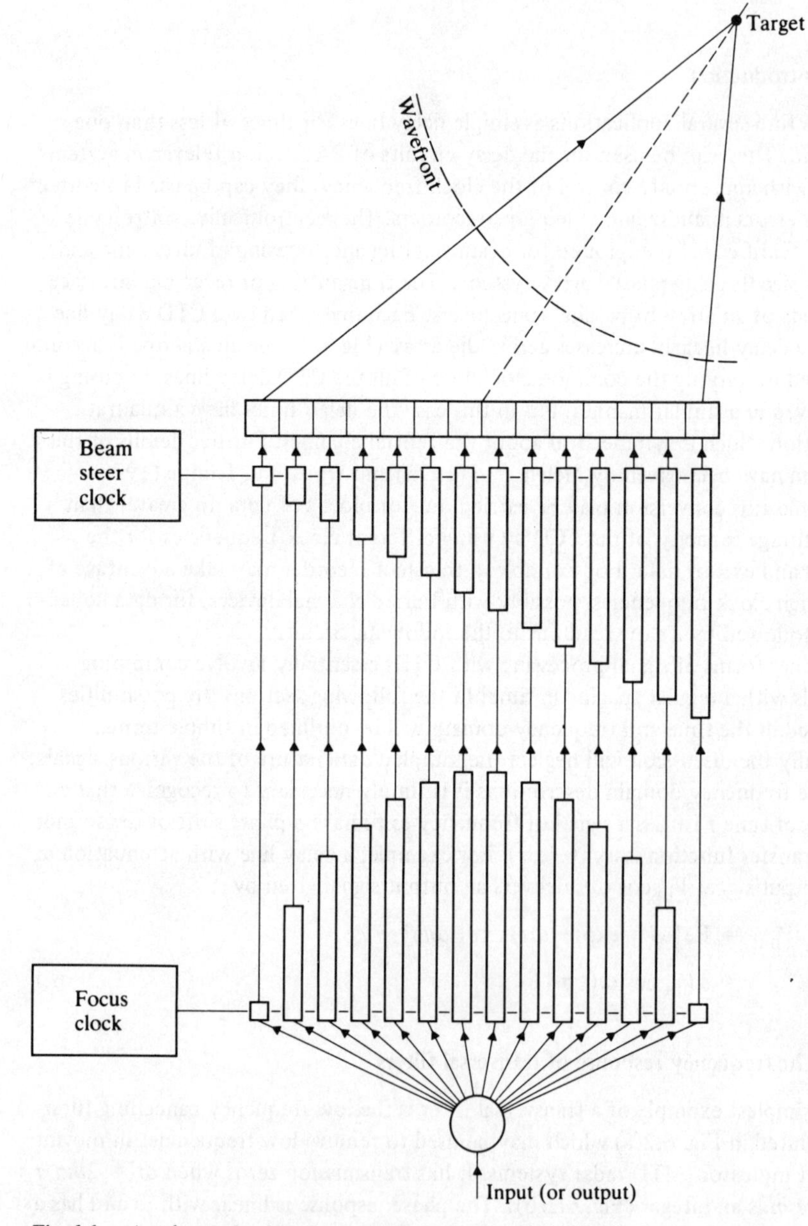

Fig. 6.1 An ultrasonic phased array using CTD to steer and focus the beam.

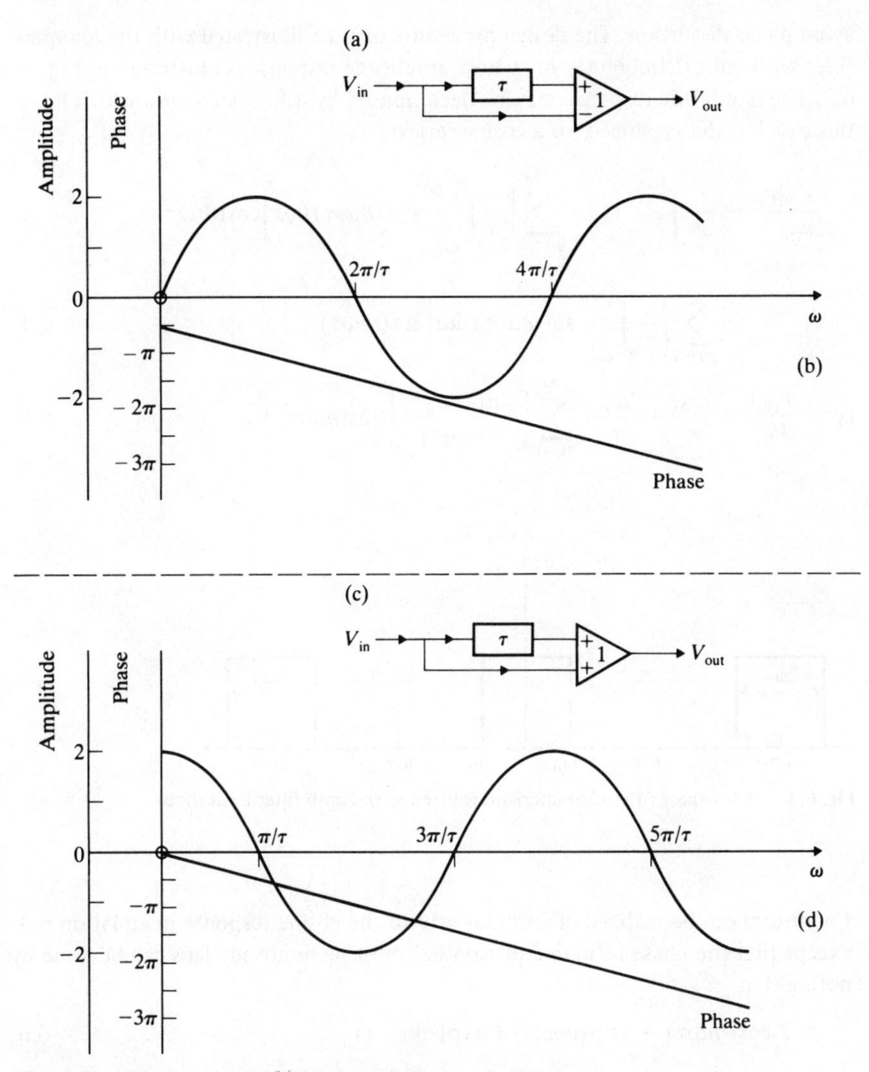

Fig. 6.2 Simple transversal filters and their characteristics.

A complementary filter is shown in Fig. 6.2(c) with the difference operation replaced by a summation. The transfer function has a linear phase characteristic which is always $\pi/2$ out of phase with the difference filter (Fig. 6.2(d)).

$$\text{i.e.}\quad V_{out}/V_{in} = \exp(-j\omega\tau) + 1$$

$$= 2\cos(\omega\tau/2)\exp(-j\omega\tau/2) \qquad 6.3$$

The comb filter behaviour in which the zeros are separated in ω by $2\omega/\tau$ is characteristic of all the delay line filters. If this feature is acceptable the amplitude transfer function can be tailored to suit a particular application while retaining the simple delay that is implied by linear phase behaviour and is desirable to avoid phase distortion. The design procedure may be illustrated with the low-pass filter with cut-off frequency ω_0 whose amplitude response is illustrated in Fig. 6.3. The amplitude response may be decomposed by a Fourier series, which in this case has the symmetry of a cosine series:

$$\frac{V_{out}}{V_{in}} = \frac{\tau}{2\pi}\int_{-\omega_0}^{+\omega_0} d\omega + \sum_{m=1}^{M}\left[\frac{\tau}{\pi}\int_{-\omega_0}^{+\omega_0} \cos(m\omega\tau)\, d\omega\right]\cos(m\omega\tau)$$

$$+ \sum_{m=1}^{M}\left[\frac{\tau}{\pi}\int_{-\omega_0}^{+\omega_0} \sin(m\omega\tau)\, d\omega\right]\sin(m\omega\tau) \qquad 6.4$$

i.e. $$\frac{V_{out}}{V_{in}} = \frac{\omega_0\tau}{\pi} + \frac{2\omega_0\tau}{\pi}\sum_{m=1}^{M}\left[\frac{\sin(m\omega_0\tau)}{(m\omega_0\tau)}\right]\cos(m\omega\tau) + 0 \qquad 6.5$$

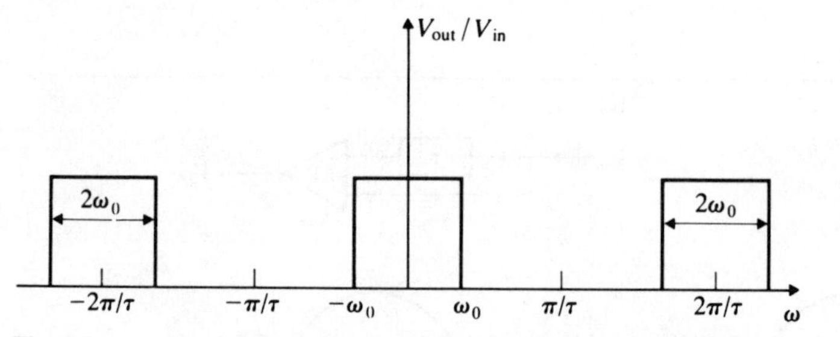

Fig. 6.3 A low-pass filter characteristic realized with comb filter limitations.

Cos$(m\omega\tau)$ can be realized in a similar way to the cosine response in equation 6.3 except that the phase term in equation 6.3 must be removed. This can be done by noting that

$$2\cos(m\omega\tau) = \exp(jm\omega\tau) + \exp(-jm\omega\tau) \qquad 6.6$$

As shown in Fig. 6.4, the individual terms can be built up with a time delay of $m\tau$ and a time advance of $m\tau$ (i.e. a delay of $-m\tau$) with each one followed by a multiplying tap weight h_m [= $(\sin m\omega_0\tau)/m\omega_0\tau$]. Exact representation of the desired pass-band shape would require an infinite number of stages but the series is truncated after M terms when a satisfactory approximation has been achieved. Unfortunately the time advance stage is a non-existent circuit element and the actual realization applies the input as shown in Fig. 6.5 so that all components whose coefficients appear as the tap weights in the triangles have an additional linear phase delay $\exp(-jM\omega\tau)$. In a CTD the tap weights would be provided through a non-destructive output technique such as the split-electrode technique in section 5.3.4 for a fixed set of tap weights. A photomicrograph of such a CCD is shown in Fig. 5.19.

An application where phase is the main interest is suggested by the frequency

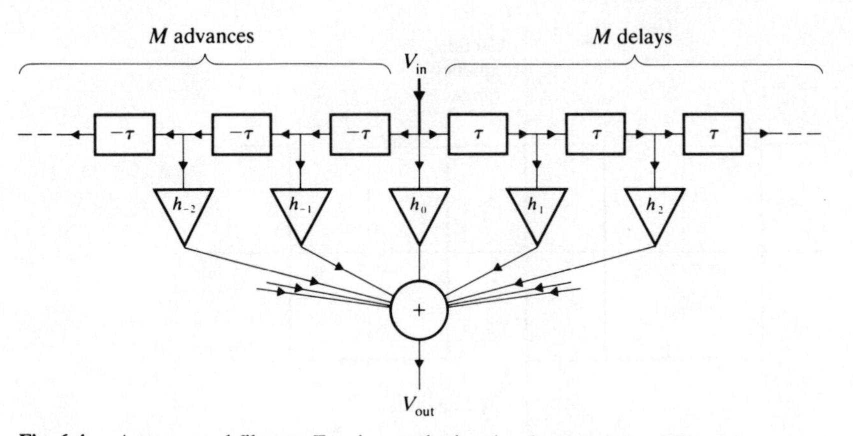

Fig. 6.4 A transversal filter to Fourier synthesize the characteristic of Fig. 6.3.

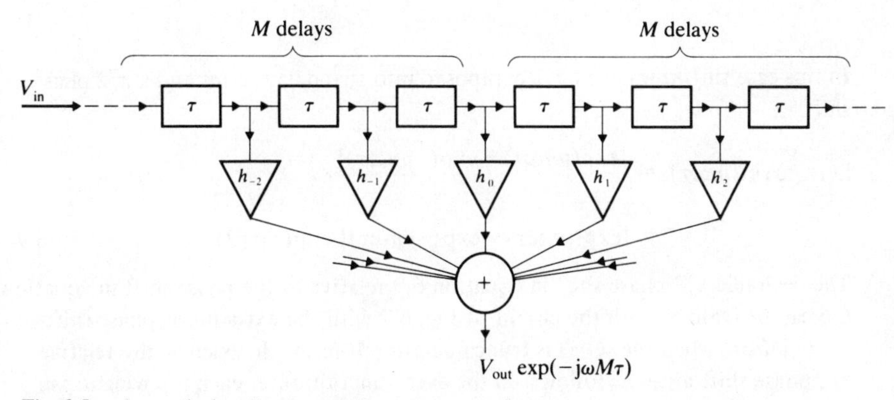

Fig. 6.5 A practical realization of the circuit in Fig. 6.4.

independent quadrature phase difference in equations 6.2 and 6.3. Such a device has use in quadraphonics, single-sideband modulation (Chowaniec and Hobson, 1976) and control systems, but requires a frequency independent amplitude response. This may be provided by the odd and even responses in Fig. 6.6. The even function can be represented by

$$\left(\frac{V_{out}}{V_{in}}\right) = 2(\cos \omega\tau - \tfrac{1}{3}\cos 3\omega\tau + \tfrac{1}{5}\cos 5\omega\tau \ldots) \qquad 6.7$$

Each term of equation 6.7 can be decomposed into two delay terms as described by equation 6.6 followed by appropriate tap weights. The odd function can be represented by

$$\left(\frac{V_{out}}{V_{in}}\right)_{odd} = 2(\sin \omega\tau + \tfrac{1}{3}\sin 3\omega\tau + \tfrac{1}{5}\sin 5\omega\tau + \ldots) \qquad 6.8$$

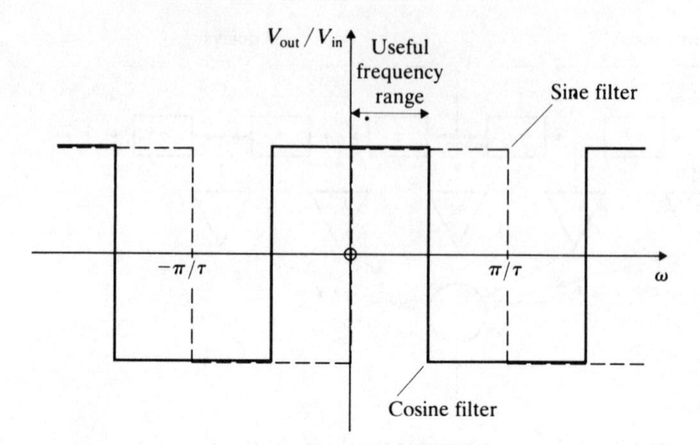

Fig. 6.6 The variation of amplitude with frequency for the two channels of a quadrature phase element.

In this case $\sin(m\omega\tau)$ can be decomposed into two delay terms and a $\pi/2$ phase shift.

$$\text{i.e.}\quad 2\sin(m\omega\tau) = \frac{[\exp(jm\omega\tau) - \exp(-jm\omega\tau)]}{j}$$

$$= [\exp(jm\omega\tau) - \exp(-jm\omega\tau)]\ \exp(-j\pi/2) \qquad 6.9$$

The desirable $\pi/2$ phase shift in equation 6.9 relative to the phase shift in equation 6.6 can be realized with the circuit of Fig. 6.7 with the extra linear phase shift $\exp(-jM\omega\tau)$ when the series is truncated after M-terms. In essence, the relative $\pi/2$ phase shift arises as follows. In the even function filter each tap-weight pair

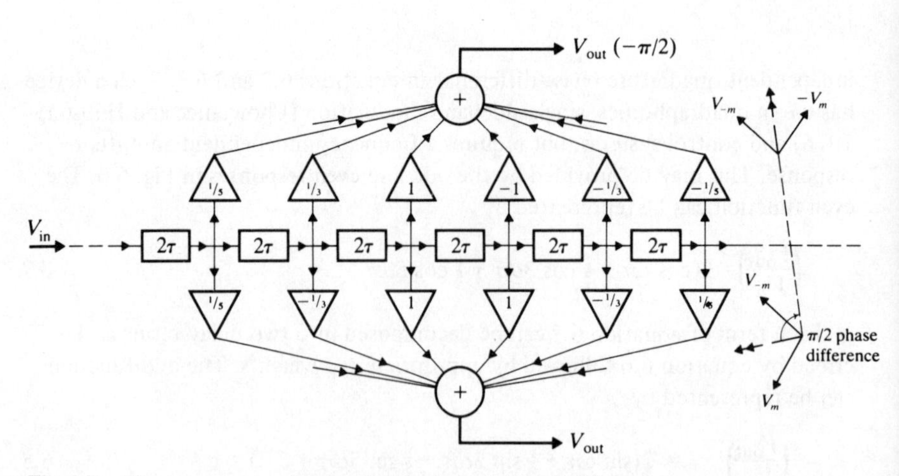

Fig. 6.7 The transversal realization of the characteristics in Fig. 6.6.

sums two signals, so providing their mean phase at the output, while the odd function filter sums the same two signals except that one of them is phase inverted. The mean phase of this pair is then in quadrature with the mean phase of the even function pair as illustrated in the insets to Fig. 6.7 for the mth tap weight pair. The quadrature phasing element has been described in terms of odd and even function tap weights for conceptual reasons. However, a simpler and more satisfactory realization of the even (cosine) function is by a single tap weight of unit amplitude at the $m = 0$ tapping point (the centre of the delay). It provides a transfer function, $\exp(-jM\omega\tau)$, to balance the same and unavoidable mean delay term of the odd function filter.

Fig. 6.6 gives the impression that only a quarter of the available frequency range $2\pi/\tau$, is used in this filter. However, Fig. 6.7 shows that the delay elements are of length 2τ so that only half the number of delay elements are required operating at half the clock frequency. In this case because even tap weights are zero the periodicity is $2\pi/2\tau$ so that half the available bandwidth is used.

The above two examples illustrated in a step by step fashion that the tap weights of a transversal filter are equivalent to the Fourier transform of the frequency response. In practice, it is more direct to recognize that the K tapping point after the input in a filter with M components in the frequency response ($2M + 1$ tapping points) has a transfer function,

$$\left(\frac{V_{out}}{V_{in}}\right)_K = h_K \exp(-jK\omega\tau)$$

$$= h_K \exp(-jM\omega\tau) \exp[j(M-K)\omega\tau] \qquad 6.10$$

In equation 6.10 all signals suffer the extra and unavoidable linear phase delay $\exp(-jM\omega\tau)$ and h_K is given by

$$h_K = \frac{\tau}{2\pi}\int_{-\pi/\tau}^{\pi/\tau} f(\omega) \exp[-j(M-K)\omega\tau]\, d\omega \qquad 6.11$$

$f(\omega)$ is the amplitude of the transfer function to be synthesized. In general h_K will be complex so requiring two taps, corresponding to the real and imaginary components, at each tapping point as in the quadrature phasing element. The factor j indicates the 90° phase shift of the odd terms. In practice $f(\omega)$ will usually be either symmetrical or antisymmetrical so that only one set of tap weights emerges as in the case of the low-pass filter.

6.3 Transversal filters in the time domain

6.3.1 Introduction to matched filtering and correlation

The tapped delay line, or transversal filter, offers the possibility of extracting signals from noise. For white additive noise the matched filter theorem (Cook and Bernfeld, 1967) states that optimum signal/noise ratio is obtained for a filter whose frequency response is the Fourier transform of the signal in the time domain. However, section 6.2 showed that the tap weights of a transversal filter are the Fourier transform of the frequency response. Therefore, the tap weights require a spatial distribution along the CTD which is the same as the shape of the signal along the time axis. There will be a peak ('correlation response') from the summation of tapped signals when the signal just 'fits in' to the CTD tapping profile. In essence, the filter attenuates strongly at those points where the signal is weak, thereby attenuating the corresponding noise, but the attenuation is small where the signal is strong. A simple example is given by a 13 bit Barker code which has the form −1, +1, −1, +1, −1, −1, +1, +1, −1, −1, −1, −1, −1, as shown in Fig. 6.8(a). The matched transversal filter is illustrated in Fig. 6.8(b). When

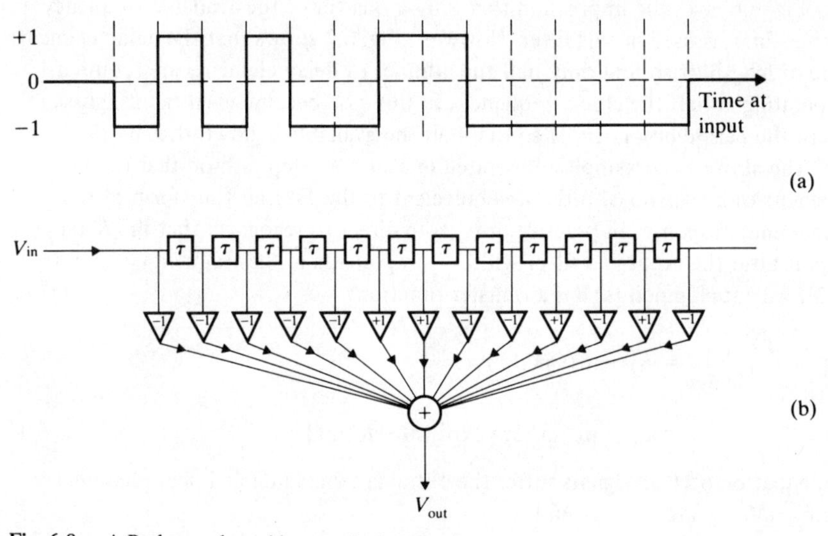

Fig. 6.8 A Barker code and its matched filter.

the code completely fills the filter a correlation peak is obtained and the output amplitude is 13 units. For one time step earlier or later the output is 0. For two steps difference the output is +1 and for three steps difference the output is 0. The noise output from each tap will only add in a root mean square fashion so that the improvement of signal/noise power ratio is $13^2/13 \simeq 11.1$ dB.

The finite response or 'side lobes' on either side of the correlation peak illustrates one problem inherent in the detection of coded signals. The dynamic range available for detection of independent and partially overlapping signals with the same code is limited to the difference of the peak and side lobes of the correlation response. Even though codes improve the signal/noise ratio they can essentially generate a further source of noise or confusion unless great care is exercised in their choice.

Another feature of correlation detection is implicit in the above description and can be seen in Figs 6.8, 6.9 and 6.10. The first 'contact' of the input signal and the correlation pattern is made by their opposite ends. If the process of signal

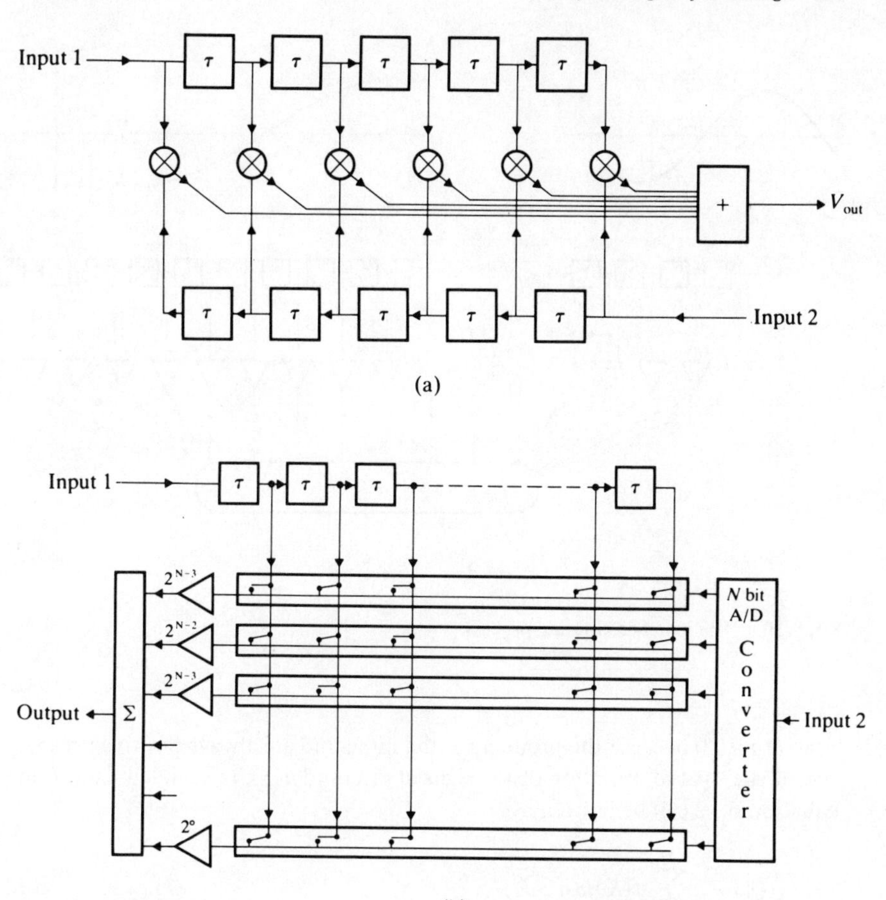

Fig. 6.9 (a) A CCD correlator; (b) an analogue–digital correlator which uses switches instead of multipliers.

flow in the CTD is viewed as two patterns running through each other then one must be the time-reversed version of the other for them to give a correlation peak as they become coincident. This feature can also be expressed in mathematical terms by the convolution of two signals $f(\lambda)$ and $h(t-\lambda)$ which is given in integral form by

$$f(t) = \int_{-\infty}^{+\infty} f(\lambda)[h(t-\lambda)]\,\mathrm{d}\lambda \qquad 6.12$$

λ is the 'spatial' coordinate of $f(\lambda)$ and $h(t-\lambda)$ written in terms of time by using the 'relative velocity' of the two signals. For the mathematically inclined reader, Cook and Bernfeld (1967) and Skolnik (1962) have given the theoretical background of convolution. If $f(\lambda)$ = constant × $h(-\lambda)$ there will be a correlation

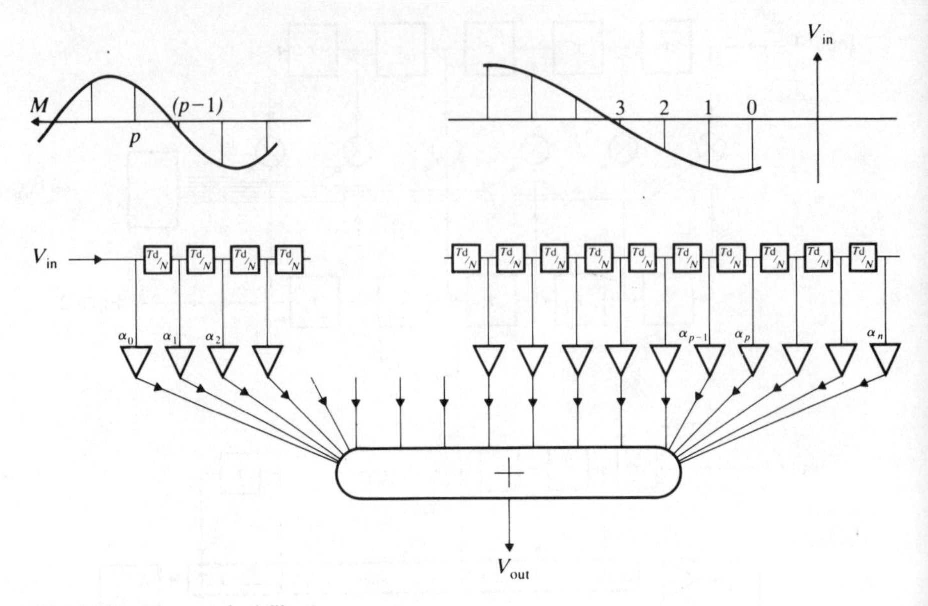

Fig. 6.10 The matched filtering process.

peak at $t = 0$ because the products in the integrand are always positive irrespective of the 'spatial' variation of the signs of $f(\lambda)$ and $h(\lambda)$. In sampled data form equation 6.12 can be written as

$$f(n) = \sum_{m=1}^{M} f(K)h(n-K) \tag{6.13}$$

n and K are integers representing units of sampled data time and there are M simultaneous comparison points between the two patterns.

If an alternative pattern recognition procedure were used in which the two waveforms to be compared were propagated in the same direction, the side lobes of the response would be much greater owing to correlation of all the parts of the signal within the detection device. There would also be a considerable timing problem, in most applications, in order to make both signals coincident.

Matched filter pattern recognition can be applied to any known shape of signal and offers a simple and potentially cheap device when fixed tap weights are required by using the split electrode technique (section 5.3.4). When programmable filtering is required the tap weights must be provided outside the CTD. The biased-gate or floating-gate tapping techniques of section 5.3.3 may be used with on-chip reset and buffering followed by a resistor connected from each tapping point to the summing point of a virtual earth amplifier. If the resistors are provided off-chip, the tap weights may be adjusted mechanically in initial setting up of the filter. Electronic control of the tap weights may be provided on- or off-chip by using a field-effect transistor as a controlled resistor. The tap weight values can be programmed and held in a non-volatile manner by using MNOS FET. Further practical details have been given by Lampe *et al* (1974).

Alternatively, the tap weighting circuits may be replaced with analogue multipliers whose outputs are all summed. Usually such a device is called a *correlator* because it takes the products of two waveforms on a point by point basis and sums the products. The second input to the analogue multipliers may be a programmable series of tap weights or it could be the tapped outputs of a second CTD which is parallel to the first CTD. The latter device would be a correlator for two time-varying signals (Fig. 6.9(a)). Practical multipliers realized with MOS devices have offset variations which presently limit the dynamic range to 40 dB for such a correlator and this is insufficient for many applications.

The problems are eased considerably if the two waveforms have a binary format because one CCD and the multipliers can be replaced by a series of binary switches controlled by a shift register. This technique can be extended for the correlation of two analogue waveforms if one of them is digitized. The purely analogue waveform is convolved with the appropriate binary digit using the series of binary switches referred to above. Each significant digit of the binary words which represent the digitized analogue waveform require a separate series of switches as shown in Fig. 6.9(b). The subsidiary correlation responses from each series of binary switches are summed with a weighting factor which decreases twofold for each significant digit. The combined response is the desired correlation response and it only requires switches rather than multipliers so that offset problems are largely removed and dynamic ranges approaching 80 dB should be possible. Unfortunately the technique is very expensive in its use of silicon chip area.

An alternative technique (Buss *et al*, 1976), is to Fourier transform the signals, multiply them and take the inverse Fourier transform but performance figures are not available.

6.3.2 Time–bandwidth product

The signal/noise improvement in the above example of detection of a Barker code was obtained by taking a signal which occupied thirteen time elements and compressing it into one time element to give the correlation peak. Noise was discriminated against much more effectively than would be the case with a simple frequency domain filtering technique because its only components to be 'recognized' were those with a temporal shape similar to that of the desired signal. This type of signal processing technique is usually characterized by the time–bandwidth product, $B\tau_c$, where B is the frequency-domain bandwidth occupied by the recognized signal and τ_c is its temporal extent. $B\tau_c$ is simply derived in a CTD. According to the sampling theorem (Shannon, 1949) the maximum frequency or bandwidth that may be reconstructed in a sampled data system has a period equal to two samples. Higher frequencies would produce a mixing product with the sampling frequency which would fall into this usable band of frequencies and would cause ambiguity or aliasing errors (see sections 5.5 and 5.6).

i.e. $$B = 1/2\tau \qquad 6.14$$

τ is the delay or storage time per stage of the CTD. If the CTD has N sample stores the signal extent that may be contained is given by

$$\tau_c = N\tau \tag{6.15}$$

From equations 6.14 and 6.15, the available $B\tau_c$ is simply half the number of CTD sample stores.

i.e. $$B\tau_c = N/2 \tag{6.16}$$

For comparison a simple second-order LCR tuned circuit has a bandwidth B_T and response or coherence time τ_T related by

$$B_T\tau_T = 1/2\pi \tag{6.17}$$

The tapped delay line clearly has a considerable power to improve the signal/noise ratio. As implied in the discussion of the Barker code, the signal power improvement is proportional to N^2 whereas the noise power increase is proportional to N so that the available signal/noise power, S/N, improvement or processing gain, g_p, for matched digital signals is

$$g_p = \frac{S/N_{out}}{S/N_{in}} = N = 2B\tau_c \tag{6.18}$$

6.3.3 Matched filters for 'chirp' waveforms

The 'chirp' waveform is often met in discussions of matched filtering. When its linear frequency modulation is superimposed on a carrier frequency, ω_0, it has the form

$$[f(t)]_{0<t<T_d} = A\cos(\omega_0 t \pm \omega_1 t^2/T_d + \phi) \tag{6.19}$$

A is the constant amplitude and the positive sign corresponds to an 'up-chirp'. The negative sign corresponds to a 'down-chirp' and T_d is the duration of the modulation if it starts at time $t = 0$. ϕ is an arbitrary phase factor. The instantaneous angular frequency, ω, is obtained by differentiating the phase in equation 6.19 with respect to time.

i.e. $$\omega = \frac{\partial}{\partial t}(\omega_0 t \pm \frac{\omega_1 t^2}{T_d} + \phi)$$

$$= \omega_0 \pm \frac{2\omega_1 t}{T_d} \tag{6.20}$$

Equation 6.20 shows that the frequency sweep bandwidth from $0 < t < T_d$ is $2\omega_1$. This waveform was originally used in pulse compression radar systems. The i.f. signal processor incorporated a dispersive delay line which allowed the end of the modulation train to catch up the leading edge just as the signals left the delay line. In this way a correlation peak was generated with signal/noise properties similar to those in equation 6.18.

In order to describe quantitatively the recognition of a waveform (such as that in equation 6.19) by a CTD match filter it is necessary to describe it in sampled data form. If the time between samples is Δt, the continuous time variable, t, is replaced by $m\Delta t$, where m is an integer which runs over the $N+1$ samples in the CTD. To be compatible with such a CTD containing $N+1$ tapping points, which encompass an overall delay of T_d provided by N delays, Δt is equal to T_d/N. Using these relationships the sampled data form of equation 6.19 is a sequence

$$A \cos\left(\frac{\omega_0 T_d}{N} m \pm \frac{\omega_1 T_d m^2}{N^2} + \phi\right) \qquad 6.21$$

where m runs from 0 to N. If this signal is fed into a tapped CTD delay line the signal output when the $p+1$ sample is at the first tapping point is

$$(V_{out})_{p \leqslant N} = A \sum_{m=0}^{m=p} h_{p-m} \cos\left(\frac{\omega_0 T_d}{N} m \pm \frac{\omega_1 T_d}{N^2} m^2 + \phi\right) \qquad 6.22$$

h_{p-m} is the $(p-m)$th tap weight reckoned from the tap nearest to the input of the CTD and is zero except in the range $0 \leqslant p-m \leqslant N$. Similarly

$$(V_{out})_{p \geqslant N} = A \sum_{m=p-N}^{m=N} h_{p-m} \cos\left(\frac{\omega_0 T_d}{N} m \pm \frac{\omega_1 T_d}{N^2} m^2 + \phi\right) \qquad 6.23$$

The relative positions of the sampled signal levels and the tap weights in the CTD is illustrated in Fig. 6.10 for $p < N$. The h_{p-m} are given values described by

$$h_q = B \cos\left(\frac{\omega_0 T_d}{N} (N-q) \pm \frac{\omega_1 T_d (N-q)^2}{N^2} + \phi\right) \qquad 6.24$$

where $0 \leqslant q \leqslant N$ and B is a constant.

After N transfers when the entire signal is contained in the CTD the two patterns are matched, so that using equations 6.23 and 6.24 the output is

$$(V_{out})_{p=N} = AB \sum_0^N \cos^2(\quad)$$

$$= \frac{AB}{2} \sum_0^N [1 + \cos 2(\quad)] \qquad 6.25$$

The empty brackets represent the angle of the cosine term in equations 6.23 and 6.24.

If the chirp signal is an integral number of periods long the cos 2( ) term in equation 6.25 sums to zero so that

$$(V_{out})_{p=N} = \frac{NAB}{2} \qquad 6.26$$

$(V_{out})_{p=N}$ is the peak voltage amplitude of the correlation response. It has been enhanced $N/2$ times compared with a simple amplitude detection of the signal and this is also the signal/noise power processing gain (compare with equation 6.18). The best performance would be obtained if the signal output for $p \neq N$

was zero, but this is not so for the above waveform. Its value can be found by evaluating the summations in equations 6.22 or 6.23 with the aid of equation 6.24 but this is somewhat tedious and will not be done here. The output signal, or side lobes, on either side of a correlation peak can be seen in Fig. 6.12 for a similar waveform which will be described below.

The chirp matched filter described by equation 6.24 is suitable when the carrier frequency, ω_0, is appreciably greater than ω_1 and the Nyquist frequency is appreciably larger than $2\omega_0$, so that the phase term, ϕ, can be absorbed as a time displacement. It is usually included in the intermediate frequency section of a receiver. Upper clock frequency limits for CTD will usually preclude this type of operation so the chirp detection must be carried out at baseband frequencies. The down-conversion process allows the phase, ϕ, to take any unknown value unless the local oscillator is phase locked to the received carrier frequency. As the latter possibility is usually inconvenient greater complexity is required in the baseband signal processing than in the intermediate frequency signal processing as will be seen in the following example and analysis.

A matched filter scheme for separately detecting up-chirps and down-chirps which chirp equally on either side of the carrier frequency is illustrated in Fig. 6.11. The chirp signal input is

$$V(t) = A\cos(\omega_0 t \pm \mu t^2 + \phi) \tag{6.27}$$

t lies within the range

$$-\frac{T_d}{2} < t < \frac{T_d}{2} \tag{6.28}$$

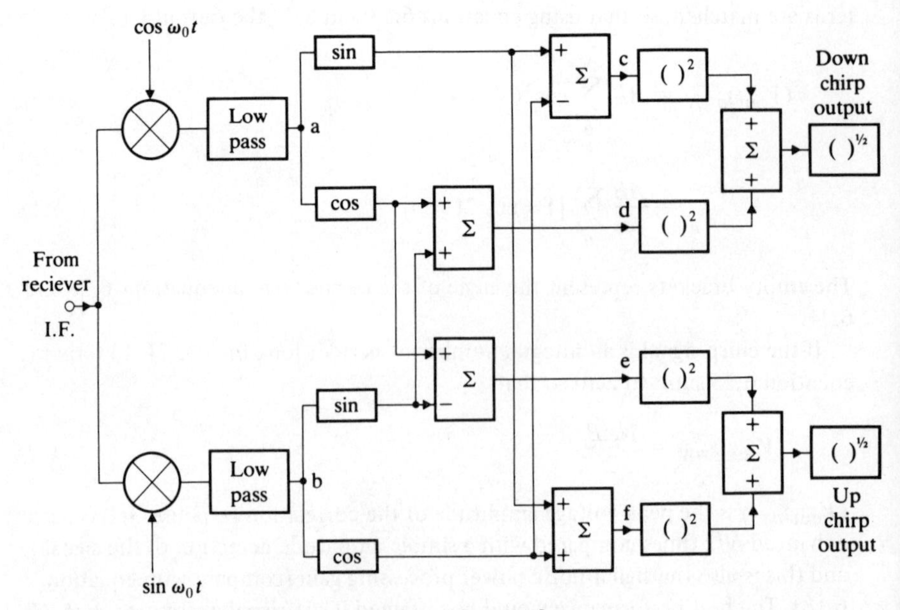

Fig. 6.11 A matched filter system to detect chirp waveforms.

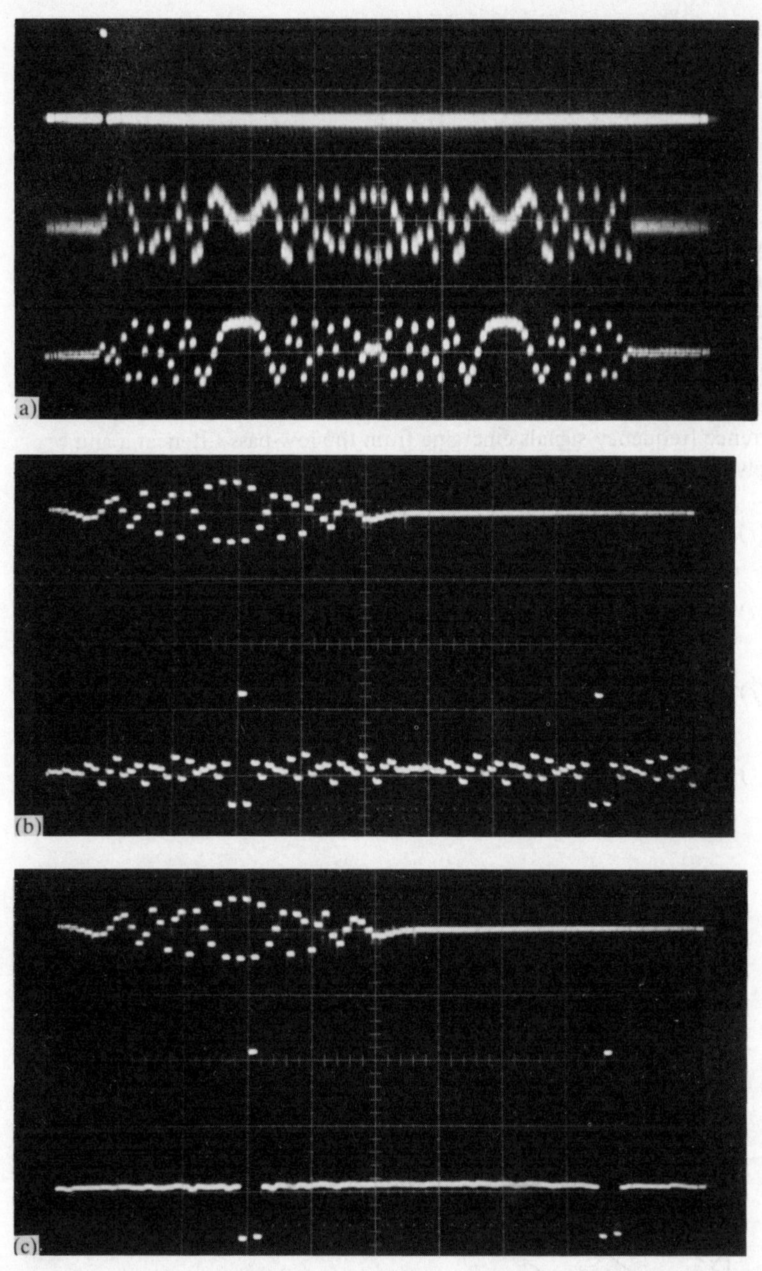

Fig. 6.12 (a) The impulse response of chirp-Z transformers. The upper trace is the impulse, the middle trace is the sine filter impulse response and the lower trace is the cosine filter impulse response. (b) The correlation response of a sine chirp-Z transformer. The upper trace is the input multiplying chirp sequence with a sine form. It also has a 40 dB Dolph–Chebyshev weighting to reduce side lobe levels. (c) As (b) except that the outputs of the sine and cosine chirp-Z transformers have been properly combined and show a considerable side lobe cancellation. The devices in these photographs have a displaced frequency origin when compared with the description in the text. The input chirp multiplying sequence runs from d.c. to twice the Nyquist frequency so that the matched filter frequencies must run from $-1 \times$ Nyquist frequency to $+3 \times$ Nyquist frequency (Wardrop and Bull, 1977). (Courtesy Marconi Research Laboratories, GEC–Marconi Electronics Ltd).

The instantaneous frequency excursion for an up-chirp, corresponding to the + sign, if from $\omega_0 - \mu T_d$ to $\omega_0 + \mu T_d$. The negative sign corresponds to a down-chirp from $\omega_0 + \mu T_d$ to $\omega_0 - \mu T_d$. When the waveforms are converted to base-band by removal of ω_0 they have the form shown in Fig. 6.12 for $\phi = 0$ and $\phi = \pi/2$. As time passes through zero, corresponding to an instantaneous chirp frequency of zero, both time and frequency change sign so both waveforms are symmetrical about their d.c. midpoint.

In equation 6.27 ϕ includes the phase difference of the input signal and local oscillator. Down-conversion by the two quadrature mixers causes sum and difference frequencies to enter the low-pass filters from the implied mixer multiplications given by $A\cos(\omega_0 t \pm \mu t^2 + \phi)\cos\omega_0 t$ and $A\cos(\omega_0 t \pm \mu t^2 + \phi)\sin\omega_0 t$. The difference frequency signals emerging from the low-pass filters at a and b respectively are

$$V_a(t) \propto \frac{A}{2}\cos(\pm\mu t^2 + \phi)$$

i.e. $$V_a(t) \propto \frac{A}{2}[\cos(\pm\mu t^2)\cos\phi - \sin(\pm\mu t^2)\sin\phi] \qquad 6.29$$

and $$V_b(t) \propto -\frac{A}{2}\sin(\pm\mu t^2 + \phi)$$

i.e. $$V_b(t) \propto -\frac{A}{2}[\sin(\pm\mu t^2)\cos\phi + \cos(\pm\mu t^2)\sin\phi] \qquad 6.30$$

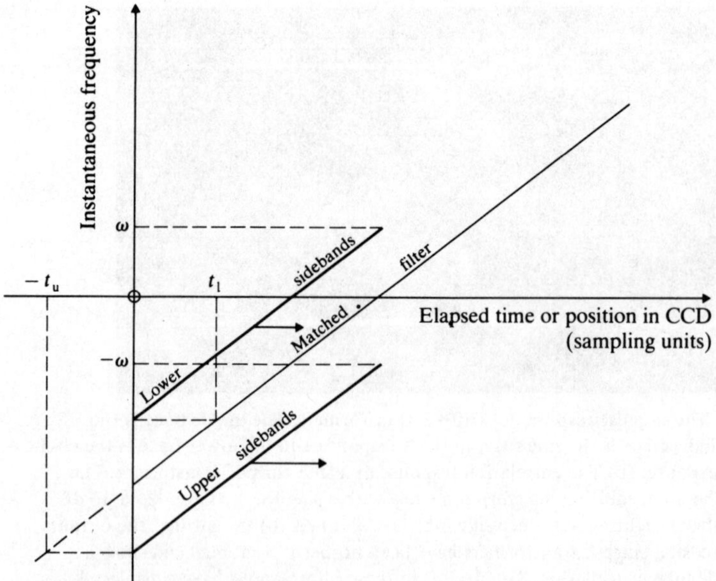

Fig. 6.13 The frequency shift caused by a single input frequency to a chirp-Z Fourier transformer.

Equations 6.29 and 6.30 are written in continuous time form rather than sampled data form for convenience. In the actual filter they would be realized in a sampled data form similar to equation 6.23. The blocks marked sin and cos in Fig. 6.11 are matched filters for sine chirps and cosine chirps so that their tap weights and impulse responses have the forms shown in Fig. 6.12. Their impulse responses are defined by

$$\text{sin: } \sin(\mu t^2) \text{ for } -T_d/2 < t < T_d/2 \qquad 6.31$$

$$\text{cos: } \cos(\mu t^2) \text{ for } -T_d/2 < t < T_d/2 \qquad 6.32$$

It should be noted that $\sin(-\mu t^2)$ will give a negative correlation peak in the sin filter.

The outputs from the matched filters each have correlation peaks at the same time and with amplitudes proportional to the product of the processing gain and the coefficients of $\sin(\mu t^2)$ and $\cos(\mu t^2)$ in equations 6.29 and 6.30 respectively so that the combined signals entering the squaring elements are

$$\left.\begin{aligned} V_c &\propto \mp\frac{A}{2}\sin\phi + \frac{A}{2}\sin\phi \\ V_d &\propto +\frac{A}{2}\cos\phi \mp \frac{A}{2}\cos\phi \\ V_e &\propto +\frac{A}{2}\cos\phi \pm \frac{A}{2}\cos\phi \\ V_f &\propto \mp\frac{A}{2}\sin\phi - \frac{A}{2}\sin\phi \end{aligned}\right\} \qquad 6.33$$

For an up-chirp the upper sign in equations 6.33 applies so that $V_c = 0$, $V_d = 0$, $V_e = A\cos\phi$, $V_f = A\sin\phi$. The output from the upper squaring, summing and square root circuit of Fig. 6.11 is then zero while that from the lower one is proportional to A. For a down-chirp the outputs are reversed. If detection of a chirp without knowledge of its direction was required only the top half or the bottom half of the circuit of Fig. 6.10 would be needed and the summing circuits following the sin and cos filters could be omitted. For the upper half the processing would then use the signal of equation 6.29 in a squaring, summing and square root sequence but the output amplitude would be halved.

6.3.4 The chirp-Z Fourier transformer

Transformation of sequential signals in the time domain to a sequence of signals equal to the amplitudes of their linearly ascending frequency components can be carried out with a modification of the chirp-Z transformer described in the previous section. In essence, the input signal is mixed with a linear chirp whose frequency rises from zero. The resulting displacement of the instantaneous frequency of the chirp causes the correlation response to occur at a shifted time which is linearly proportional to the frequency of the incoming signal. The process is illustrated in Fig. 6.13 for an input frequency ω.

To express the process analytically we start from the Fourier transform

$$f(\omega) = \frac{2}{T_d}\int_{T_1}^{T_1+T_d} f(t)\exp(-j\omega t)\,dt \qquad 6.34$$

T_1 is the starting time for data acquisition in the time T_d. The positive real and negative imaginary parts of $f(\omega)$ in equation 6.34 are the amplitudes of the cosine and sine components of the signal at frequency ω. Equation 6.34 can be converted to sampled data form in the same way that equation 6.19 was converted to equation 6.21.

$$\text{i.e.}\quad f(\omega) = \frac{2}{N}\sum_{0}^{N-1} f_n \exp\left(-j\omega T_d \frac{n}{N}\right) \qquad 6.35$$

f_n is the sequence of data points that constitute $f(t)$. The integral in equation 6.34 is independent of T_1 if an integral number of periods, $2\pi/\omega$, occupy the time, T_d. The evaluated frequencies can then be described by an integer, K.

$$\text{i.e.}\quad \omega = K\,2\pi/T_d \qquad 6.36$$

Substituting equation 6.36 in equation 6.35,

$$F_K = \frac{2}{N}\sum_{0}^{N-1} f_n \exp\left(-j2\pi K \frac{n}{N}\right) \qquad 6.37$$

F_K is the sequence of data points which are the amplitudes corresponding to frequencies $(2\pi/T_d)K$. Equation 6.37 is known as the discrete Fourier transform. The maximum value of K is $N/2$ corresponding to Nyquist limited sampling in f_n.

The chirp-Z transform can be obtained from the discrete Fourier transform by using the identity

$$-2nK = (n-K)^2 - n^2 - K^2 \qquad 6.38$$

$$\text{i.e.}\quad F_K = \exp\left(-\frac{j\pi K^2}{N}\right)\sum_{n=0}^{N-1}\left\{\left[f_n \exp\left(-\frac{j\pi n^2}{N}\right)\right]\exp\left(\frac{j\pi(n-K)^2}{N}\right)\right\} \qquad 6.39$$

The product $f_n \exp(-j\pi n^2/N)$ can be implemented at the input to the CTD processor by multiplying the data sequence, f_n, by the chirp sequence, $\exp(-j\pi n^2/N)$, which is held in a suitable memory. The next procedure is to carry out the convolution of this chirp multiplied input data sequence with a chirp matched filter. The two phases contained by each of the complex exponentials in equation 6.39 implies that parallel multiplying and parallel matched filtering are required as illustrated in Fig. 6.14. For an input frequency with index K corresponding to an angular frequency $2\pi K/T_d$, the correlation peak from the matched filters occurs K samples after the chirp multiplied input sequence has completely entered the CTD.

The maximum value of K in equation 6.39 is $N/2$ because this limit cor-

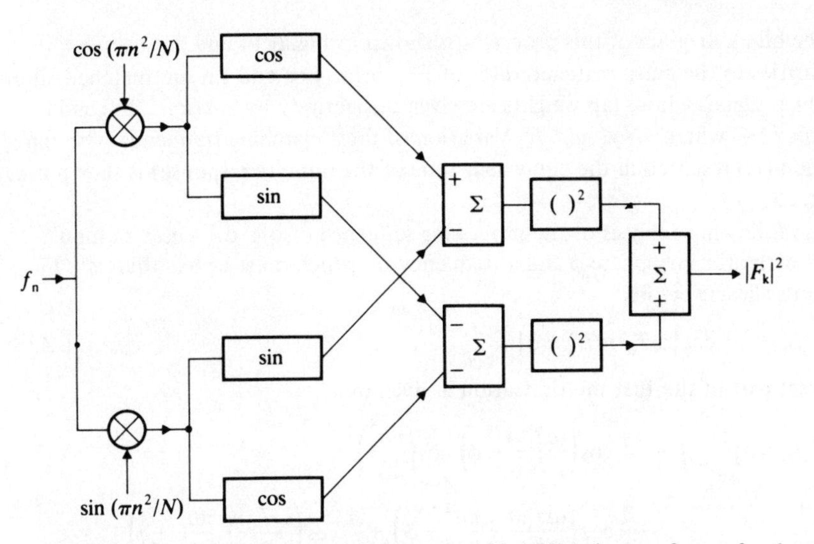

Fig. 6.14 The circuits required to implement a chirp-Z Fourier transformer for the power spectral density of a signal.

responds to Nyquist limited sampling in f_n. Also the term $\exp(-j\pi n^2/N)$ exceeds the Nyquist limited pattern of two samples per cycle for $n > N/2$.

A more useful form of chirp-Z transform with twice the number of output points is obtained if K is defined by

$$\omega = K(\pi/T_d) \qquad 6.40$$

The Fourier transform defined by equation 6.34 is still independent of T_1 for values of ω defined by equation 6.40. Substituting equation 6.40 in equation 6.35 and using equation 6.38 the chirp-Z transform is

$$F_K = \exp\left(-\frac{j\pi K^2}{2N}\right) \sum_{n=0}^{N-1} \left[f_n \exp\left(-\frac{j\pi n^2}{2N}\right) \exp\left(\frac{j\pi(n-K)^2}{2N}\right)\right] \qquad 6.41$$

In equation 6.41 the frequency in the exponentials becomes equal to the Nyquist limit when the index is equal to N and there are N transform points, F_K, in the output for N input data points in f_n.

If the phase information of the Fourier transform must be preserved, the data sequence in the parallel output channels must be multiplied by $\exp(-j\pi K^2/2N)$. Details have been given by Mayer (1975). Such a course would be necessary if the Fourier transformation was followed by the reverse transformation after filtering had been carried out in the frequency domain. A simpler requirement for most cases is the power density spectrum of the input signal given by $F_K F_K^*$. This is the only one to be considered here and is given by

$$F_K F_K^* = \left\{ \sum_{n=0}^{N-1} \left[\left(f_n \exp\left(-\frac{j\pi n^2}{2N}\right)\right) \exp\left(\frac{j\pi(n-K)^2}{2N}\right)\right]\right\}^2 \qquad 6.42$$

The block diagram of this processor is shown in Fig. 6.14 and has many similarities to the chirp matched filter of Fig. 6.11. Cos and sin are matched filters for chirp signals whose tap weights are given respectively by $\cos(\pi m^2/2N)$ and $\sin(\pi m^2/2N)$ where $-N < m < N$. Variation of their matching frequency with tap position (represented in the equivalent time of the impulse response) is shown in Fig. 6.13.

The following analysis of the processing sequence of Fig. 6.14 uses an input signal with arbitrary phase ϕ and a frequency ω, which must be less than $\pi N/T_d$ to avoid aliasing errors.

i.e. $$f_n = A\cos[\omega T_d(n/N) + \phi] \tag{6.43}$$

The real part of the first multiplication is given by

$$\begin{aligned} f_n \cos\left(\frac{\pi n^2}{2N}\right) &= A\cos\left(\frac{\omega T_d n}{N} + \phi\right)\cos\left(\frac{\pi n^2}{2N}\right) \\ &= \frac{A}{2}\cos\left(\frac{\omega T_d n}{N} + \frac{\pi n^2}{2N} + \phi\right) + \frac{A}{2}\cos\left(\frac{\omega T_d n}{N} - \frac{\pi n^2}{2N} + \phi\right) \\ &= \frac{A}{2}\cos\left(\frac{\omega T_d n}{N} + \frac{\pi n^2}{2N}\right)\cos\phi - \frac{A}{2}\sin\left(\frac{\omega T_d n}{N} + \frac{\pi n^2}{2N}\right)\sin\phi \\ &\quad + \frac{A}{2}\cos\left(\frac{\omega T_d n}{N} - \frac{\pi n^2}{2N}\right)\cos\phi - \frac{A}{2}\sin\left(\frac{\omega T_d n}{N} - \frac{\pi n^2}{2N}\right)\sin\phi \end{aligned} \tag{6.44}$$

The upper sideband terms involving $\omega T_d n/N + \pi n^2/2N$ have a linear down-chirp from the point of furthest penetration into the CTD at a frequency $-\omega$. The terms involving $\omega T_d n/N - \pi n^2/2N$ down-chirp through zero instantaneous frequency at a time displacement $t_l (= \omega T_d^2/\pi N)$ from the point of furthest penetration into the CTD. It is these lower sideband terms which can be detected with the matched filters as indicated diagramatically in Fig. 6.13. The correlation peak occurs after a time t_l, which corresponds to a frequency index K where $K = \omega T_d/\pi$. If ω is not integrally related to π/T_d the correlation peak will be shared between two adjacent Ks (i.e. it will emerge in two successive output time intervals).

In a similar fashion to the signal separation carried out in equation 6.44 the imaginary part of the first multiplication is given by

$$\begin{aligned} f_n \sin\left(\frac{\pi n^2}{2N}\right) &= \frac{A}{2}\sin\left(\frac{\omega T_d n}{N} + \frac{\pi n^2}{2N}\right)\cos\phi + \frac{A}{2}\cos\left(\frac{\omega T_d n}{N} + \frac{\pi n^2}{2N}\right)\sin\phi \\ &\quad - \frac{A}{2}\sin\left(\frac{\omega T_d n}{N} - \frac{\pi n^2}{2N}\right)\cos\phi - \frac{A}{2}\cos\left(\frac{\omega T_d n}{N} - \frac{\pi n^2}{2N}\right)\sin\phi \end{aligned} \tag{6.45}$$

As before, the signals recognized in the sin and cos matched filters are the ones containing $\omega T_d n/N - \pi n^2/2N$.

The signal amplitude entering the upper summation of Fig. 6.14 is

$$V_{S_1} \propto \frac{A}{2}\cos\phi + \frac{A}{2}\cos\phi = A\cos\phi \tag{6.46}$$

The signal amplitude entering the lower summation is

$$V_{S_2} \propto \frac{A}{2} \sin \phi + \frac{A}{2} \sin \phi \;=\; A \sin \phi \qquad 6.47$$

Applying the remaining squaring and adding to V_{S_1} and V_{S_2} shows that the output is proportional to the power spectral density, A^2. It occurs $K(= \omega T_d/\pi)$ samples after the N signal samples have entered the matched filters as discussed above.

Reference to Fig. 6.13 shows that the upper sidebands would have a subsidiary correlation peak at time $-t_u$, even though only part of the input sequence had entered the CTD. For this reason the output circuit must be blanked during the period of the N data entries to the CTD. The input frequency must be within the extreme frequencies of the chirp signals for correlation to occur and reference to Fig. 6.13 indicates that the highest possible input frequency would require N stages beyond and including the frequency zero for correlation to occur. The CTD has to have $2N - 1$ stores. The central one giving a zero frequency tap weight is at opposite ends of the active tap weights for zero and maximum frequencies when correlation occurs. Fig. 6.13 also implies that the input must be blanked after the N multiplier signal samples have entered the CTD so that additional samples are not allowed to interfere with the correlation process. Summarizing the above disadvantages, the input and output blanking only allow the CTD to be used for 50% duty cycle on continuous signals and the CTD has to have $2N - 1$ stores in order to process N data samples.

6.3.5 The sliding chirp-Z Fourier transformer

The two defects of the chirp-Z transformer are overcome neatly by the sliding chirp-Z transformer. The chirp waveform used in the input multiplication chirps up and down symmetrically about zero frequency. Its extreme instantaneous frequencies are the Nyquist frequencies so that frequencies other than zero in the input signal cause aliasing as shown in Fig. 6.15. The chirp matched filters also chirp symmetrically between Nyquist frequency extremes and have the same number of points. For a zero input frequency there will be a correlation peak each time that the multiplying chirp has reached one extreme frequency so that the signal in the matched filter is coincident with the tap weight pattern. If the input signal contains a finite frequency it will generate upper and lower sidebands on the multiplying chirp as shown in Fig. 6.15. There will now be two signal patterns which can generate correlation responses in the matched filter and they do so at times symmetrically displaced about the time for the zero frequency correlation response. Neither of these signal patterns is purely an upper sideband or a lower sideband. Each one has a component of the other at one of its extremes in order to make up the complete linear chirp pattern. The extra piece of the pattern is generated by the alias process which 'folds' any sideband components in excess of the Nyquist frequency symmetrically about the Nyquist frequency. Care must be taken in ensuring continuity of the phase at the multiplying chirp reversal points so that the junction of the upper and lower sideband

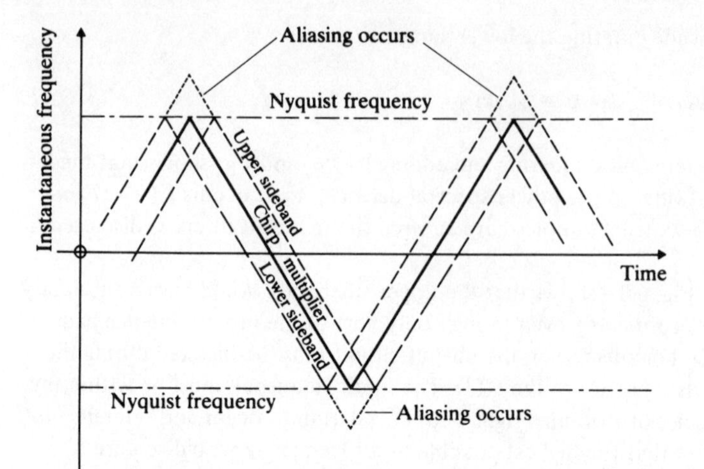

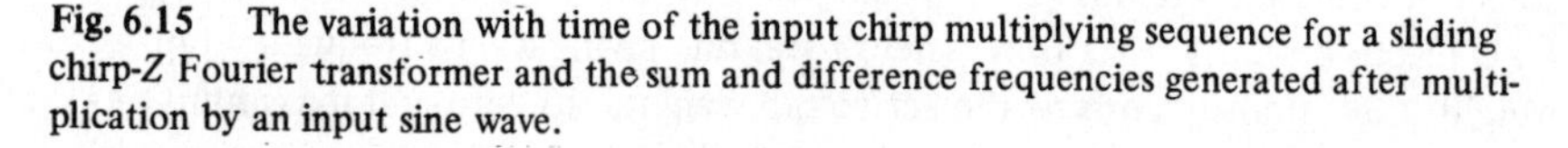
Fig. 6.15 **The variation with time of the input chirp multiplying sequence for a sliding chirp-*Z* Fourier transformer and the sum and difference frequencies generated after multiplication by an input sine wave.**

components occurs as a smooth waveform compatible with the matched filter. The time of the correlation peak is linearly related to the input frequency, with the input multiplier timing as a reference, just as it was for the chirp-*Z* transformer discussed in the previous section. The process takes place continuously and the CTD need only have as many signal stores as there are multiplying points. The formal details of the sliding CZT are similar to those of the conventional CZT and are given by Bailey *et al* (1975) and Buss *et al* (1976).

The amplitude of each frequency is derived with a different set of data so that phase information is corrupted. This arises from the incremental time shift between adjacent frequency outputs during which one piece of data is lost from the end of the CTD and another enters at the input. In the case of a periodic signal, the corruption of the phase information is irrelevant if the power density spectrum is the only requirement as in the processor of Fig. 6.14. If the form of the signal is time varying the power density spectrum will be accurate providing that the changes occur over a time long compared with $2N-1$ data samples, where N is the number of tapping points in the matched filters. If the waveform is white noise the processing will again be accurate owing to the stationary nature of the power density spectrum (Bell, 1960).

An advantage of both Fourier transformers lies in their use of matched filters which will enhance the signal/noise ratio of the detection process in proportion to the number of tapping points, as discussed in sections 6.3.1 and 6.3.2. The dynamic range of the transformer will be limited by the unwanted side lobe responses unless care is taken to suppress them. Inaccurate tap weighting could cause problems in this respect but a more serious problem can arise from the finite data length of the transformation which superimposes a $(\sin x)/x$ response on the desired response. This is reduced by modifying the tap weights near the end of the matched filter in order to make the data sequence appear to have a greater length. The technique is known as apodization and one sequence of tap weight modifications is the Hamming weighting. Further comments on the errors in this type of Fourier transformer and a comparison with the digital fast fourier transformer have been given by Buss *et al* (1975).

6.3.6 Waveform generators

As has been remarked in earlier sections, the impulse response of a transversal matched filter has the same form as the tap weights of the CTD. The impulse response is realized by propagating a single charge pulse along the CTD. In this way arbitrary waveforms of finite length are easily generated at the summing point of the tap weights. The amplitude is controlled by amplitude of the impulse and the rate may be controlled with the clock frequency.

6.4 Recursive filters in the frequency domain

6.4.1 The simple recursive filter

In contrast with the feedforward nature of transversal filters the output signal of a recursive filter is fed back to the input so that many circulations may take place. The simplest example is illustrated in Fig. 6.16 where the feedback loop contains a delay τ and a buffer amplifier with gain A. In the steady state,

$$V_{out} = V_{in} + A\,e^{-j\omega\tau}\,V_{out}$$

Writing V_{out}/V_{in} as the transfer function $H(\omega)$,

$$H(\omega) = V_{out}/V_{in} = \frac{1}{1 - A\,e^{-j\omega\tau}} \qquad 6.48$$

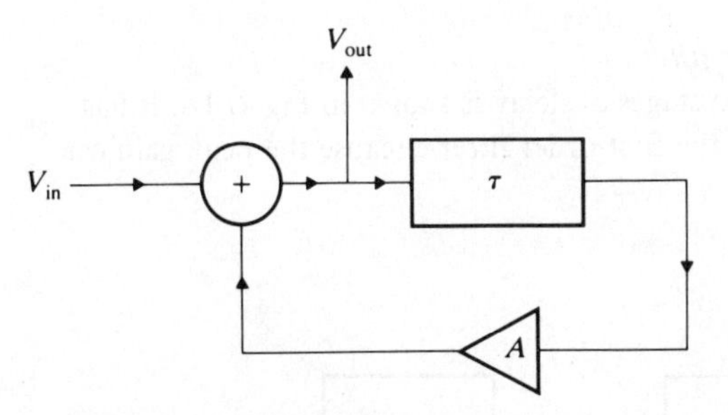

Fig. 6.16 A simple recursive filter.

A, including any reduction of amplitude in the CTD, must be less than unity to avoid oscillations. The amplitude and phase characteristics can be seen by re-arranging equation 6.48 to give

$$H(\omega) = \frac{\exp(j\omega\tau/2)\exp\left\{j\arctan\left[\left(\frac{1+A}{1-A}\right)\tan\frac{\omega\tau}{2}\right]\right\}}{(1 - 2A\cos\omega\tau + A^2)^{\frac{1}{2}}} \qquad 6.49$$

They are illustrated in Fig. 6.17 for positive A where the comb response characteristic of a delay line filter can be seen. The linear phase characteristic, described by the phase factor $\exp(j\omega\tau/2)$, is only obtained for $A = -1$ when it passes through zero phase at zero frequency or for $A = +1$ when it passes through phase $\pi/2$ at zero frequency. The amplitude transfer function becomes infinite for $\omega\tau = (2n+1)\pi$, where n is an integer and $A = 1$, and it is infinite for $\omega\tau = 2n\pi$ for $A = +1$. For $|A| < 1$ the peaks are finite with a gain $1/(1 - |A|)$.

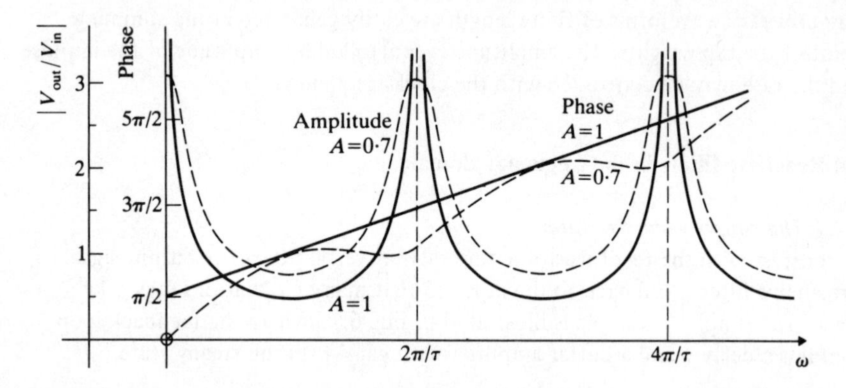

Fig. 6.17 The transfer function of a simple recursive filter.

In this case the relationship of phase and frequency is not linear and has a $\pi/2$ deviation at the frequencies of peak gain.

6.4.2 The second-order recursive filter

The second-order filter using two stages of delay is shown in Fig. 6.18. It has more interesting behaviour than the first-order filter because the peak gain can

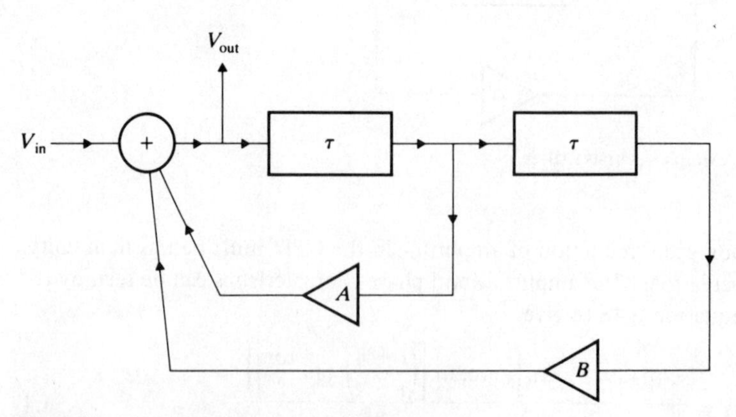

Fig. 6.18 A second-order recursive filter.

be made to occur at frequencies which are not integrally related to the inverse of the delay time. It is a basic building block of many more complicated filters. The following analysis is made in simple terms for the reader not familiar with Z-transform techniques. The Z-transform analysis has been given by Smith *et al* (1972). If an input signal $V_{in} \exp(j\omega t)$ is applied to the circuit of Fig. 6.18 the output signal is given by

$$V_{out}(t) = A V_{out}(t-\tau) + B V_{out}(t-2\tau) + V_{in} \exp(j\omega t) \qquad 6.50$$

The time in parentheses is the time at which V_{out} is to be evaluated. Equation 6.50 may be expected to have solutions of the form

$$V_{out}(t) = P \exp(st) + H(\omega) V_{in} \exp(j\omega t) \qquad 6.51$$

$H(\omega)$ is complex and is the desired transfer function for steady state operation. Any undesirable oscillations will grow from some fluctuation P and their initial small signal growth is described by the complex frequency s. If s is explicitly written as

$$s = \sigma + j\omega \qquad 6.52$$

positive σ will describe growth of an oscillatory signal with frequency ω. The boundary between stability and instability will be $\sigma = 0$.

Substituting equation 6.51 in equation 6.50 and rearranging gives

$$P \exp[s(t-2\tau)] [\exp(2s\tau) - A \exp(s\tau) - B]$$
$$+ V_{in} \exp[j\omega(t-2\tau] \{H(\omega)[\exp(2j\omega\tau) - A \exp(j\omega\tau) - B] - \exp(2j\omega\tau)\}$$
$$= 0 \qquad 6.53$$

The instability term containing P and the desirable response containing V_{in} have independent time variations so that their corresponding terms in equation 6.53 must be separately equal to zero. The instability is determined by

$$\exp(2s\tau) - A \exp(s\tau) - B = 0 \qquad 6.54$$

i.e. $$2 \exp(s\tau) = A \pm (A^2 + 4B)^{\frac{1}{2}} \qquad 6.55$$

Using equation 6.52 in equation 6.55,

$$2e^{\sigma\tau}(\cos \omega\tau + j \sin \omega\tau) = A \pm (A^2 + 4B)^{\frac{1}{2}} \qquad 6.56$$

The solutions of equation 6.56 have two distinct regions depending on the sign of $A^2 + 4B$. If $A^2 + 4B > 0$, the right hand side of equation 6.56 is purely real. Therefore the instability frequencies have a simple inverse relationship to the delay time given by

$$\sin \omega\tau = 0 \qquad 6.57$$

$$\therefore \quad \cos \omega\tau = \pm 1$$

The growth rate σ is given by

$$(\sigma\tau)_{A^2+4B>0} = \ln \left(\frac{|A \pm (A^2 + 4B)|}{2} \right) \qquad 6.58$$

The boundary between stability and instability is given by $\sigma = 0$.

i.e. $|A \pm (A^2 + 4B)^{\frac{1}{2}}| = 2$ 6.59

If $A^2 + 4B < 0$, which requires B negative, equation 6.56 may be written as

$$2e^{\sigma\tau}(\cos \omega\tau + \mathrm{j} \sin \omega\tau) = A \pm \mathrm{j}(4|B| - A^2)^{\frac{1}{2}} \qquad 6.60$$

σ is found by equating the modulus of each side of equation 6.60.

i.e. $(\sigma\tau)_{A^2+4B<0} = \ln(|B|)^{\frac{1}{2}}$ 6.61

In this case the boundary between stability and instability is given by $|B| = 1$.

i.e. $B = -1$ 6.62

The instability frequencies may be obtained from equation 6.60.

i.e. $\tan \omega\tau = \pm(4|B| - A^2)^{\frac{1}{2}}/A$ 6.63

Equation 6.63 shows that these resonant frequencies do not have a simple inverse relationship with the delay τ but are also determined by A and B. The values of A and B allowing stable operation are shown by the shaded triangle in Fig. 6.19.

Having identified the stable region attention may be usefully turned to the transfer function which from equation 6.53 is given by

$$H(\omega) = \exp(2\mathrm{j}\omega\tau)/(\exp(2\mathrm{j}\omega\tau) - A\exp(\mathrm{j}\omega\tau) - B) \qquad 6.64$$

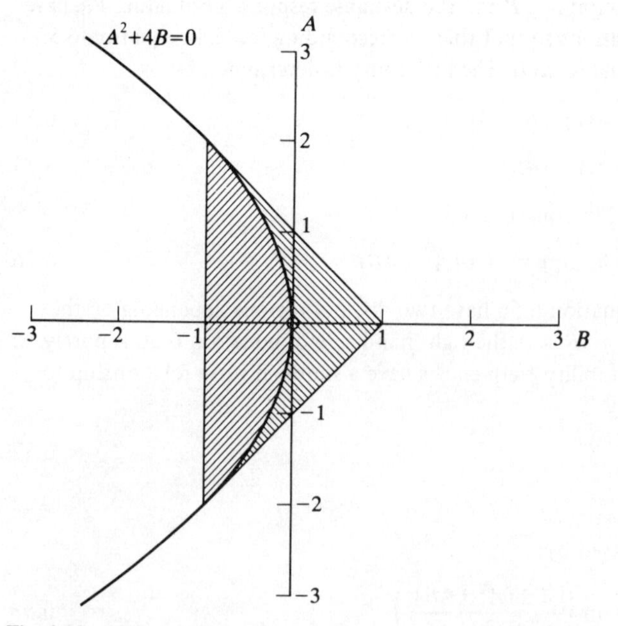

Fig. 6.19 The stability region of a second-order recursive filter.

The points of greatest interest are the amplitude peaks or resonances. It will be assumed that these coincide with the real frequency component of the poles of equation 6.64 when $j\omega$ is replaced by the complex frequency variable s. Conveniently, the poles are given by equation 6.54. For the condition $A^2 + 4B > 0$, equation 6.57 shows that the poles have the same frequencies as those of the simple recursive filter. They occur at frequencies which are odd or even multiples of $1/2\tau$ depending on the sign of $A \pm (A^2 + 4B)^{\frac{1}{2}}$. The second-order nature can make the resonant responses occupy a narrower bandwidth than those of the simple recursive filter. This can be seen on the boundary $A^2 + 4B = 0$ when equation 6.64 simplifies to

$$[H(\omega)]_{A^2+4B=0} = 1/[1 - (A/2)\exp(-j\omega\tau)]^2 \qquad 6.65$$

Equation 6.65 is essentially the squared version of equation 6.48 and has an infinitely high peak for $A = 2$.

More interesting behaviour occurs for $A^2 + 4B < 0$. Equation 6.61 shows that the transient decay rate of an oscillatory signal is only controlled by one feedback factor, B. This may be related to the 3 dB bandwidth of the circuit (at resonance frequencies given by equation 6.63) as follows. A signal, $V(t)$, at the resonant frequency ω_r, has its decay described by

$$V(t) \propto \exp(\sigma t)\exp(j\omega_r t) \qquad 6.66$$

The energy stored in the circuit is

$$\epsilon(t) \propto |V(t)|^2 \qquad 6.67$$

and the energy dissipated per cycle is

$$\Delta\epsilon(t) \propto |V(t)|^2 - |V(t + 2\pi/\omega_r)|^2 \qquad 6.68$$

The circuit Q-factor is

$$Q = 2\pi \times \text{energy stored/energy dissipated per cycle} = 2\pi\epsilon(t)/\Delta\epsilon(t) \qquad 6.69$$

i.e. $$Q = 2\pi/[1 - \exp(4\pi\sigma/\omega_r)] \qquad 6.70$$

For high Q circuits, $\sigma \ll \omega$,

$$\therefore \quad Q \simeq -\omega_r/2\sigma = \omega_r/2|\sigma| \qquad 6.71$$

Q may be written as $\omega_r/2\pi BW$ for a high-Q circuit where BW is the full 3 dB bandwidth of the resonance (Van Valkenburg, 1955).

$$\therefore \quad BW = |\sigma|/\pi \qquad 6.72$$

Returning to the second-order filter and substituting equation 6.61 in 6.72,

$$(BW)_{A^2+4B<0} = \ln(|B|)/2\pi\tau \qquad 6.73$$

Equation 6.73 implies the possibility of electronic control of a resonance bandwidth by variation of B alone. The resonant frequencies may be calculated from equation 6.63 which may be written in the simpler form

$$\cos \omega_r\tau = A/2|B|^{\frac{1}{2}} \tag{6.74}$$

For constant B, the resonant frequencies as shown in Fig. 6.20 move from even multiples of π/τ for A positive and $A^2 + 4B = 0$ through odd multiples of $\pi/2\tau$ for $A = 0$ to odd multiples of π/τ for A negative and $A^2 + 4B = 0$. While this feature offers the promise of electronic control of the resonant frequency by control of A it is more convenient in practice to vary τ by varying the clock frequency. In Fig. 6.20 the arrows indicate the variation of the resonant frequency with decreasing A.

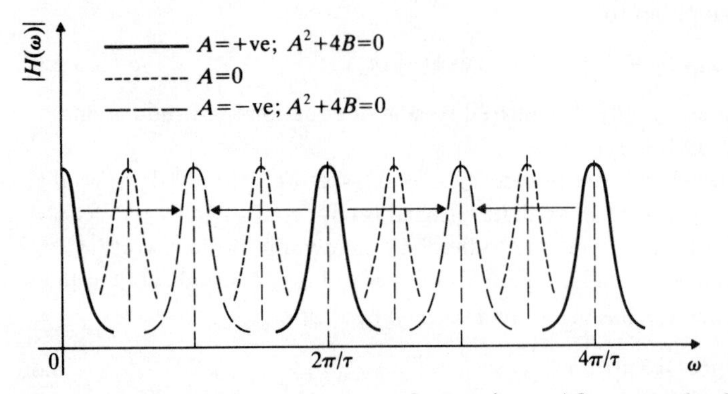

Fig. 6.20 The dependence of resonant frequencies on A for a second-order recursive filter.

If the resonant frequency had been calculated from the maximum magnitude of $H(\omega)$ in equation 6.64, rather than from the real frequency part of the complex pole frequencies, they would have been given by

$$(\cos \omega_r\tau)_{\max|H(\omega)|} = A(1 + |B|)/4|B| \tag{6.75}$$

In the high-Q limit as $|B| \to 1$,

$$A(1 + |B|)/4|B| \simeq A/2|B|^{\frac{1}{2}} \tag{6.76}$$

Equation 6.76 indicates that equations 6.74 and 6.75 are equivalent for most practical situations.

6.5 Recursive filters in the time domain – video integrators

The simple recursive filter may be used to integrate repetitive signals as illustrated in Fig. 6.21. If the time between successive samples is equal to the delay time of the CTD the recirculated and incident pulses will add coherently. Any noise present on the input signal has random phase and amplitude so that it adds in a root mean square manner. In this way the signal/noise power ratio increases with the number of integrated pulses. After m circulations with a loop gain G the peak output signal is given by

$$V_{out}/V_{in} = 1 + G + G^2 + G^3 + \ldots + G^m$$
$$= (1 - G^{m+1})/(1 - G) \tag{6.77}$$

In continuous operation $G < 1$ to avoid spurious oscillations. As $m \to \infty$,

$$(V_{out}/V_{in})_{m=\infty} = 1/1 - G \tag{6.78}$$

Equation 6.78 could have been derived directly from the steady state operation of the circuit which requires that

$$V_{out} = V_{in} + V_{out}\, G \tag{6.79}$$

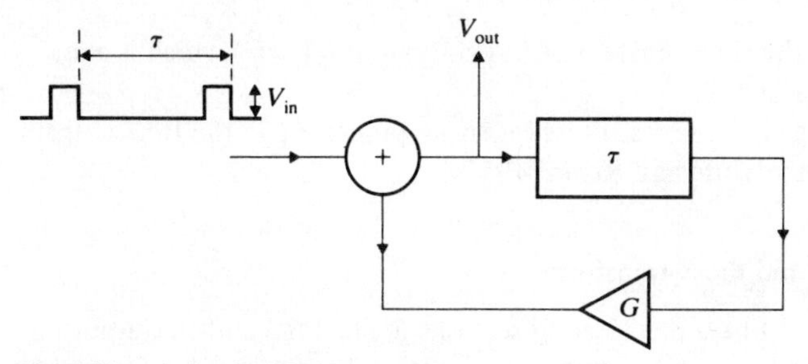

Fig. 6.21 A simple video integrator.

If the r.m.s. noise input voltage is $V_{n,in}$, the r.m.s. noise output voltage, $V_{n,out}$, after m circulations is given by

$$V_{n,out}/V_{n,in} = (1 + G^2 + G^4 + \ldots + G^{2m})^{\frac{1}{2}}$$

$$= [(1 - G^{2m+2})/(1 - G^2)]^{\frac{1}{2}} \tag{6.80}$$

In the limit $m \to \infty$ for $G < 1$,

$$V_{n,out}/V_{n,in} = 1/(1 - G^2)^{\frac{1}{2}} \tag{6.81}$$

The signal/noise power ratio improvement can be obtained from equations 6.77 and 6.80

$$\left[\frac{(S/N)_{out}}{(S/N)_{in}}\right]_m = \frac{(1 - G^{m+1})(1 + G)}{(1 + G^{m+1})(1 - G)} \tag{6.82}$$

As $m \to \infty$,

$$\frac{(S/N)_{out}}{(S/N)_{in}} = \frac{(1 + G)}{(1 - G)} \tag{6.83}$$

The above equation indicates that the output signal and the signal/noise ratio improvement tend to infinity as $G \to 1$. Such a condition has two defects. The infinite improvement requires an infinite number of circulations and so is impractical. Also, imperfect practical circuit mountings do not allow G to be frequency independent so G has to be less than unity to avoid spurious oscillations at any frequency. In order to deal with the first defect we need to know the time to carry out the greater part of the integration. This can be obtained from the rate of change of V_{out} as follows

$$\Delta V_{out}/\Delta t = [(V_{out})_{m+1} - (V_{out})_m]/\tau$$

$$\text{i.e.} \quad \Delta V_{out}/\Delta t = [V_{in} - (1-G)(V_{out})_m]/\tau \qquad 6.84$$

On integrating equation 6.84 and requiring the output to be zero at $t = 0$, we have

$$(V_{out})_m = V_{in}(1 - \exp[-t(1-G)/\tau])/(1-G) \qquad 6.85$$

From equation 6.85 the time constant of integration is $\tau/(1-G)$ which is equal to $1/(1-G)$ recirculations.

Applications using higher-order filters for signal processing in the time domain have been described by White and Ruvin (1957).

6.6 Canonical filters and the *Z*-transform

Each of the transversal filters described in sections 6.2 and 6.3 had an output which was derived from a sequence of input signals while the recursive filters described in sections 6.4 and 6.5 had an output derived from a sequence of previous output signals. The transversal filters had zeros in their frequency response while the recursive filters had poles. It seems natural that a combination of both techniques should be most versatile for general filter design. Furthermore, conventional lumped circuit filter synthesis, which has produced Butterworth, Chebyshev and Elliptic function responses, is specified in terms of poles and zeros so it should be possible to take over such design techniques and use them in low-frequency applications where conventional circuits are cumbersome and expensive. Before these techniques of synthesis are discussed further it is necessary to introduce some of the convenient analytical terminology.

If the nth output sample is represented by $y(n\tau)$ and it is determined by the previous m values of the output (an mth order recursive connection) and the previous r values of the input (an rth-order transversal connection), the circuit equation relating input and output is

$$y(n\tau) = \sum_{i=0}^{r-1} L_i x(n\tau - i\tau) + \sum_{i=1}^{m} K_i y(n\tau - i\tau) \qquad (6.86)$$

L_i and K_i are the appropriate tap weights, and τ is the delay time between samples. The transversal and recursive connections could be made with separate tapped delay lines but the most economical circuit in terms of component usage is illustrated in Fig. 6.22. Even so, Fig. 6.22 is not necessarily the preferred circuit because its efficiency may cause too great a dependence of the performance on one component. Examples of such problems have been given by White and Ruvin (1957). $\omega(n\tau)$ in Fig. 6.22 is an intermediate signal which is the output from all the recursive stages and is effectively the input to the transversal stages.

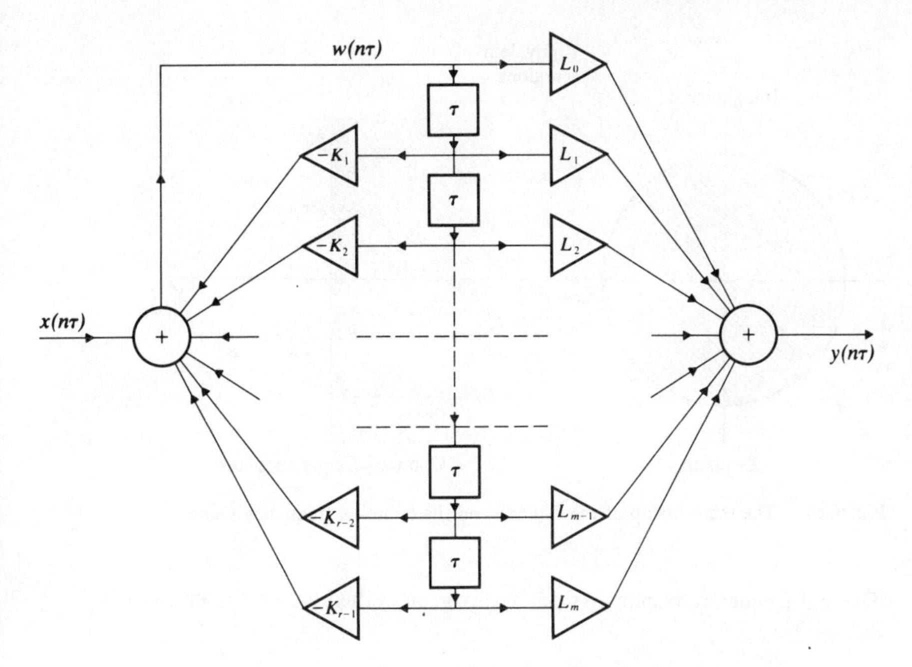

Fig. 6.22 A canonical filter configuration of recursive and transversal connections.

The frequency response of a circuit described generally by equation 6.86 has the characteristic comb filter response which repeats at frequency intervals of $2\pi/\tau$. The role of the Laplace transform in this type of circuit is as important as in lumped circuit analysis but becomes more cumbersome in the complex frequency plane owing to the repeated frequency character. In sampled data form the term $\exp(st)$, where s is the complex frequency of the Laplace transform, becomes $\exp(ns\tau)$. It is then more convenient to replace $\exp(s\tau)$ by a complex number Z. In the same way, the transfer function of a delay, τ, in Laplace transform terms is $\exp(-s\tau)$ which becomes Z^{-1}.

Instead of describing the details of a circuit, such as its poles and zeros, in the complex frequency plane the description uses the Z-plane. If a frequency domain response (i.e. steady state) is required, Z in any expression is replaced by $\exp(\mathrm{j}\omega\tau)$ just as s is replaced by $\mathrm{j}\omega$ when there is a requirement for the steady state response which is a limit of the more general complex frequency description. Two features of the Z-plane representation may be obtained from this discussion. In the left-hand half of the complex frequency plane (see Fig. 6.23 where σ is the real part of s and ω is its imaginary part) the modulus of the complex frequency is less than unity so that the corresponding region of the Z-plane is defined by

$$|Z| = |\exp(s\tau)| = e^{\sigma\tau} < 1 \qquad 6.87$$

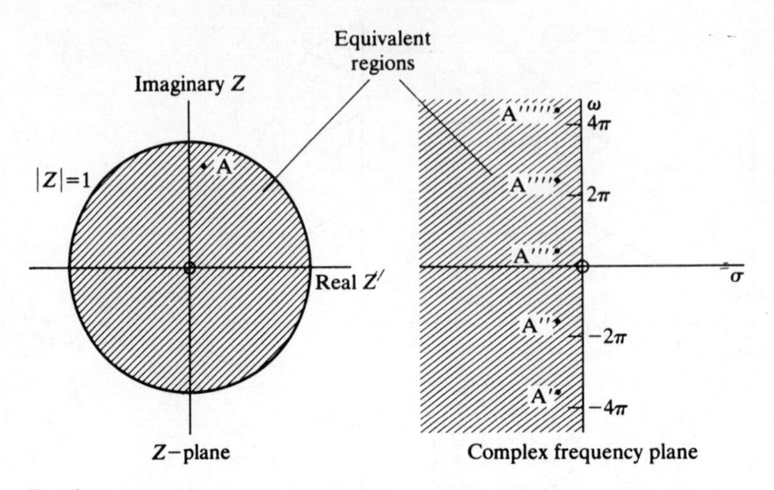

Fig. 6.23 The relationship of the *Z*-plane and the complex frequency plane.

The real frequency response (s purely imaginary) then lies on the unit circle defined by

$$|Z| = |\exp j\omega\tau| = 1 \qquad 6.88$$

If the real frequency is increased from ω to $\omega + 2\pi/\tau$, which is the repeat distance along the frequency axis of a comb filter response, Z is unchanged. This continuous frequency change corresponds to a tracing out of one revolution around a circle in the Z-plane. In this way an infinite number of poles or zeros ($A_1{}'\ A_1{}''$ etc.) spaced by 2π in ω in the complex frequency plane become one point (A) in the Z-plane as shown in Fig. 6.23.

In the above discussion the transfer function of a delay, τ, was shown to be Z^{-1}. If this is applied to equation 6.86,

$$x(n\tau - i\tau) = Z^{-i}x(n\tau)$$

and $$y(n\tau - i\tau) = Z^{-i}y(n\tau) \qquad 6.89$$

Equation 6.86 can be rearranged with the aid of equations 6.89 to give the transfer function, $H(Z)$, of the circuit of Fig. 6.22.

i.e. $$H(Z) = \frac{y(n\tau)}{x(n\tau)} = \frac{\sum_{i=0}^{r-1} L_i Z^{-i}}{1 - \sum_{i=1}^{m} K_i Z^{-i}} \qquad 6.90$$

Equation 6.90 shows that $H(Z)$ can be written as the ratio of two rational polynomials and it implies that the circuit of Fig. 6.22 can be used to realize any Z-plane transfer function that can be written as the ratio of two polynomials. Accordingly, the circuit of Fig. 6.22 is usually called a canonical filter.

The form of equation 6.90 is the same as the form of the transfer function in the complex frequency plane of continuous filters arising from techniques of network synthesis except that there the variable is s. This feature suggests that the vast array of knowledge of network synthesis can be applied to delay line filters of canonical or other configurations. The filters realized in this way cannot be exactly the same as continuous filters. The latter have a transfer function which is specified continuously and non-repetitively up to infinite frequency while the former have a repeating frequency range of $2\pi/\tau$ with a symmetry in shape at every π/τ. In effect the 'infinite' frequency of delay line filters is the Nyquist frequency π/τ. Recognizing this fact, the continuous frequency filter shapes can be modified by a suitable non-linear frequency scaling of the form

$$\omega/\omega_0 \to C \tan(\omega\tau/2) \tag{6.91}$$

In equation 6.91 ω_0 is a cut-off frequency of the continuous filter while C is a scaling constant for the delay line filter necessary to place its cut-off frequency at any desired frequency. It should also be noted that the frequency scaling between the two types of filter is approximately linear as $\omega \to 0$. Equation 6.91 can be written as

$$\frac{\omega}{\omega_0} \to -\mathrm{j}C \frac{\exp(\mathrm{j}\omega\tau) - 1}{\exp(\mathrm{j}\omega\tau) + 1} \tag{6.92}$$

In complex frequency terms equation 6.92 can be written as

$$s/\omega_0 \to C(Z - 1)/(Z + 1) \tag{6.93}$$

In order to use existing knowledge of circuit synthesis it is simply necessary to identify, in the complex frequency plane, the poles and zeros of the transfer function of the continuous filter which is used as the starting point of the design. The poles or zeros are then transformed using equation 6.93 to the Z-plane giving points Z_{p1}, Z_{p2}, Z_{p3}, etc. and Z_{01}, Z_{02}, Z_{03}, etc., respectively. The transfer function of the desired delay line filter is then given by

$$H(Z) = \frac{(Z - Z_{01})(Z - Z_{02})(Z - Z_{03}) \ldots}{(Z - Z_{p1})(Z - Z_{p2})(Z - Z_{p3}) \ldots} \tag{6.94}$$

Equation 6.94 can be factored to give the desirable recursive and transversal connections and their tap weights. More details and specific examples have been given by White and Ruvin (1957) and Gold and Rader (1969).

An example of a third-order Chebyshev design realized in sampled data form is illustrated in Fig. 6.24. The objective is to remove stationary and low-speed clutter from a MTI radar signal (Roberts *et al*, 1973) and it represents a considerable improvement on the simple cancelling filter of Fig. 6.2. In essence the first delay stage is a simple transversal canceller with very little recursive feedback. The second and third delay stages are a second-order recursive filter as described in section 6.4.2 (with $A^2 + 4B < 0$) combined with a second-order transversal canceller. The overall frequency response is illustrated in Fig. 6.25.

In addition to the above properties the Z-representation can be manipulated

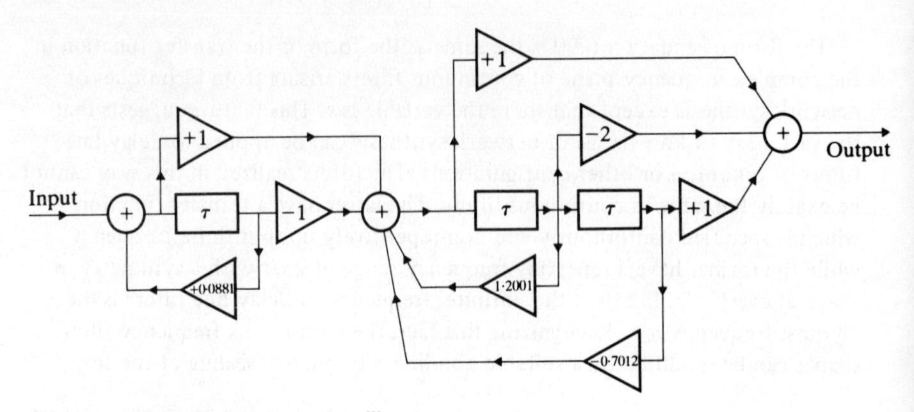

Fig. 6.24 A third-order Chebyshev filter.

by a Z-transform which has the same manipulative value as the Laplace or Fourier transforms. It is given by

$$X(Z) = \sum_{n=0}^{\infty} x(n\tau)Z^{-n} \qquad 6.95$$

In this case $x = 0$ for $n < 0$. For example the response of cascaded circuits, whose Z-transforms are separately known, to a signal whose Z-transform is also known is obtained by simply multiplying their Z-transforms just as the response of cascaded circuits in the frequency domain can be obtained by taking their Fourier transform. An example of this use is given in section 6.7.

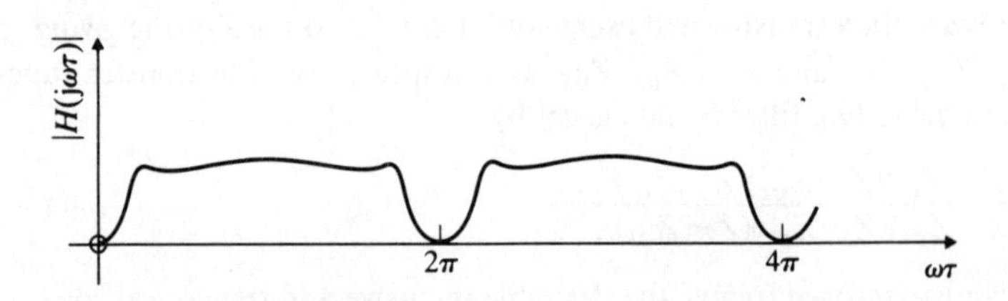

Fig. 6.25 The amplitude response of the third-order Chebyshev filter in Fig. 6.24.

If it is desired to recover a temporal waveform the inverse Z-transform (Rader and Gold, 1967; White and Ruvin, 1957; Gold and Rader, 1969) is used and is given by

$$x(n\tau) = \frac{1}{2\pi j}\oint X(Z)Z^{n-1}\,dZ \qquad 6.96$$

In this case the integration path must enclose all the poles in the Z-plane.

6.7 Errors in transversal and recursive filters

Imperfect charge transfer and inaccuracies of tap weighting are sources of error in charge-transfer devices. In the frequency domain, imperfect charge transfer causes both an amplitude error owing to the signal attenuation and a phase error owing to the additional delay suffered by smeared charge. These effects have been quantitatively described in equations 4.20 and 4.21. The effect of these errors on a transversal filter is to limit the depth of the nulls and to shift their frequency. The former defect will also arise from tap weighting inaccuracies. The effects can be quantitatively illustrated for the simple transversal filter of Fig. 6.26 where α contains both the attenuation error and the direct-path tap

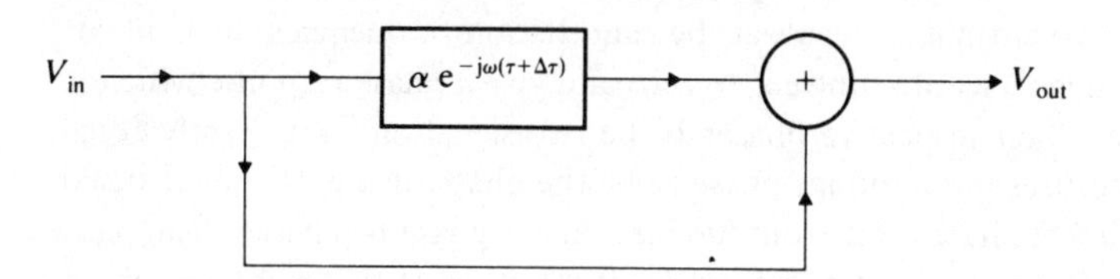

Fig. 6.26 Error terms in a simple transversal filter.

weighting error. The effect on the amplitude of the transfer function is illustrated in Fig. 6.27 (Chowaniec and Hobson, 1976). The measured α was also dependent on frequency in this example. When the phase response is considered the most serious defect arises from the amplitude unbalance in the parallel feedforward

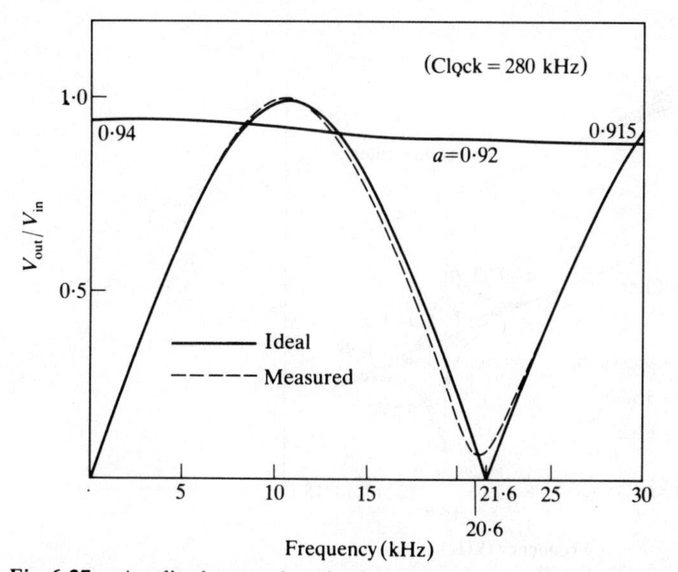

Fig. 6.27 Amplitude errors in a simple transversal filter. There is also a smearing induced shift in the frequency for minimum response.

signal channels. Neglecting the phase shift caused by charge smearing, which simply shifts the null frequencies, the transfer function of the circuit in Fig. 6.26 is

$$V_{out}/V_{in} = 1 + \alpha \exp(-j\omega\tau) \tag{6.97}$$

Equation 6.97 may be rewritten as

$$\frac{V_{out}}{V_{in}} = \cos\frac{\omega\tau}{2}\exp\left(-\frac{j\omega\tau}{2}\right)\left\{(1+\alpha)\left[1 + j\frac{(1-\alpha)}{(1+\alpha)}\tan\frac{\omega\tau}{2}\right]\right\} \tag{6.98}$$

The terms outside the square brackets are the ideal response. At the cancellation points where $\omega\tau/2$ is an odd multiple of $\pi/2$ the phase error is $\pi/2$ for *any* deviation of α from unity. The physical explanation of this defect arises from the addition of the two antiphase signals at the cancellation frequencies. In an ideal system the two signals would subtract to zero and give a phase response which is half way between their respective phases. If the two signals are not exactly equal the subtraction causes the resultant phase to be the phase of the dominant signal so that there is a $\pi/2$ phase error from the ideal linear phase behaviour. The phase lags that of the ideal response if $\alpha > 1$ but leads if $\alpha < 1$. At first sight this phase error is not important because the amplitude response is small. However, in the region around the nulls the phase error remains large as shown for both a summing and a difference filter in Fig. 6.28. This restricts the useful bandwidth of

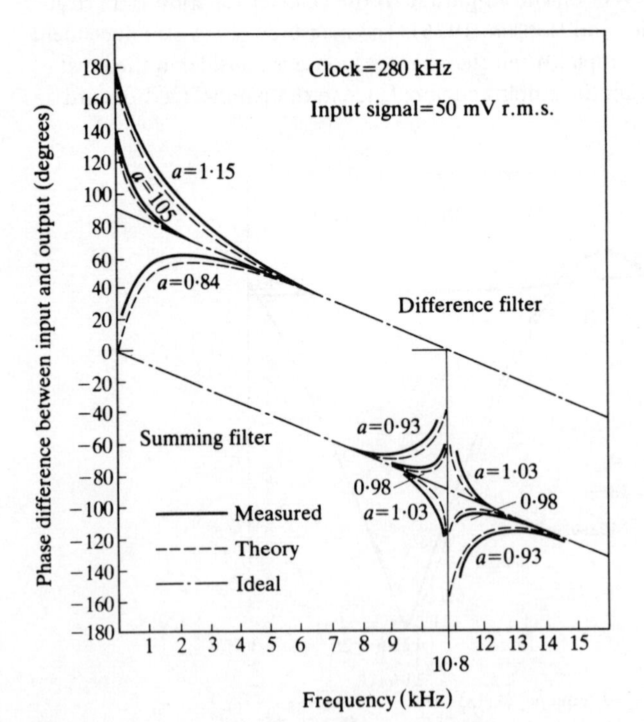

Fig. 6.28 Phase errors in a simple transversal filter caused by unequal amplitudes in each path.

any filter which requires phase accuracy. Quantitative limitations in a 90° phasing element have been discussed by Chowaniec and Hobson (1976).

The above considerations apply to the more general case of practical transversal filters with many signal paths. Computational convenience is obtained by describing the effects with the Z-transform as shown in the following development. At the first transfer a relative quantity of charge $(1-\epsilon)$ is transferred and ϵ remains. The second transfer is $(1-\epsilon)\epsilon$ and ϵ^2 remains. This process continues so that the transfer generates the following infinite sequence of charge-transfer events from an initial unit pulse

$$(1-\epsilon); (1-\epsilon)\epsilon; (1-\epsilon)\epsilon^2; (1-\epsilon)\epsilon^3; \ldots \textit{ad infinitum}$$

From equation 6.95 the Z-transform of this sequence is

$$X(Z) = \sum_{n=1}^{\infty} (1-\epsilon)\epsilon^{n-1}Z^{-n}$$

$$= Z^{-1}(1-\epsilon)/(1-\epsilon Z^{-1}) \qquad 6.99$$

In equation 6.99 the summation starts at $n = 1$ because the first quantity of charge, $1-\epsilon$, only 'passes the observation point' after one transfer event. In a CTD with N stages in cascade the overall Z-transform, $X_N(Z)$, is the product of the Z-transforms of the individual stage. If ϵ is assumed to be the same at each stage,

$$X_N(Z) = \left[\frac{Z^{-1}(1-\epsilon)}{(1-\epsilon Z^{-1})}\right]^N \qquad 6.100$$

$X_N(Z)$ is now the appropriate quantity to multiply by the tap weight after N stages. Equation 6.100 shows that the Z-transform is simply Z^{-N} (corresponding to a perfect delay of N stages) if $\epsilon = 0$. For many practical cases where $\epsilon \ll 1$ equation 6.100 can be approximated in the following more convenient form

$$X_N(Z) = Z^{-N} \exp\{N[\ln(1-\epsilon) - \ln(1-\epsilon Z^{-1})]\}$$

$$= Z^{-N} \exp[N\epsilon(Z^{-1}-1)] \qquad 6.101$$

A similar expression has already been met in section 4.6.

In a transversal matched filter we are concerned with the summation of many expressions of the form of equation 6.101 for different signal paths.

i.e. $$H(Z) = \sum h_N X_N(Z) X_S(Z) \qquad 6.102$$

h_N is the tap weight and $X_S(Z)$ is the Z-transform of the input signal. We cannot proceed beyond equation 6.102 without considering numerical examples for h_N and $X_S(Z)$. In principle modifications to h_N could be made to correct for the incomplete charge-transfer defects in $X_N(Z)$ but this is usually impractical owing to lack of knowledge of ϵ when the tap weights h_N are defined in the production process. However, a matched filter is an optimum filter so that any errors can only decrease the output signal. As this comment is true for positive as well as negative errors the overall error cannot be a first-order effect. Therefore matched filters are immune to errors to first-order (Buss *et al* 1974).

Charge smearing can cause serious errors when there is a time division multiplexing of signals in the CTD. One example is the video integrator discussed in section 6.5. Each of the integrated pulses in Fig. 6.21 only occupies one storage element of the CTD. In a radar system successive storage elements of the CTD would contain information about targets in a sequence of finite range elements (range 'bins'). Charge smearing will limit the dynamic range of the device when a large signal and a small signal occupy adjacent range bins and will generate false target information. The flowchart of Fig. 6.29 illustrates how the spurious secondary and tertiary signals build up in successive circulations of the CTD for

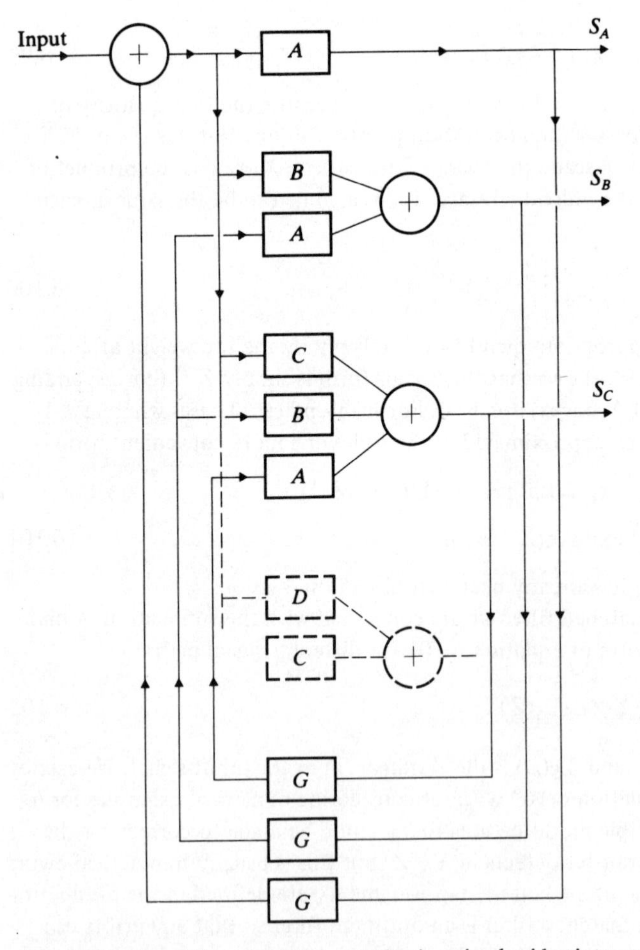

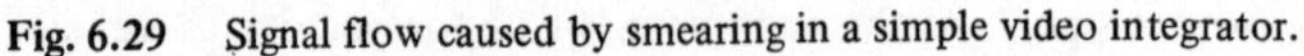

Fig. 6.29 Signal flow caused by smearing in a simple video integrator.

a repetitive input of unit amplitude. A, B and C are the magnitudes of the successive components of a unit impulse response and are given respectively by $(1-n\epsilon)$, $n\epsilon(1-n\epsilon)$, $(n\epsilon)^2(1-n\epsilon)/2$ assuming $n \gg 1$ and $n\epsilon \ll 1$. After m circulations the primary, secondary and tertiary output signals are, respectively,

$$S_A = A[1 + GA + (GA)^2 + (GA)^3 + \ldots + (GA)^{m-1}]$$
$$S_B = B[1 + 2GA + 3(GA)^2 + \ldots + m(GA)^{m-1}]$$
$$S_C = C[1 + 2GA + 3(GA)^2 + \ldots + m(GA)^{m-1}] + GB^2\,[0 + 1 + 3GA + 6(GA)^2 + \ldots + m(m-1)(GA)^{m-2}/2] \qquad 6.103$$

G is the loop gain excluding the transfer inside the CTD. For positive values of G any one spurious output builds up faster than the ones leading it because it feeds on all the (growing) leading outputs as well as on the input signal. As $m \to \infty$,

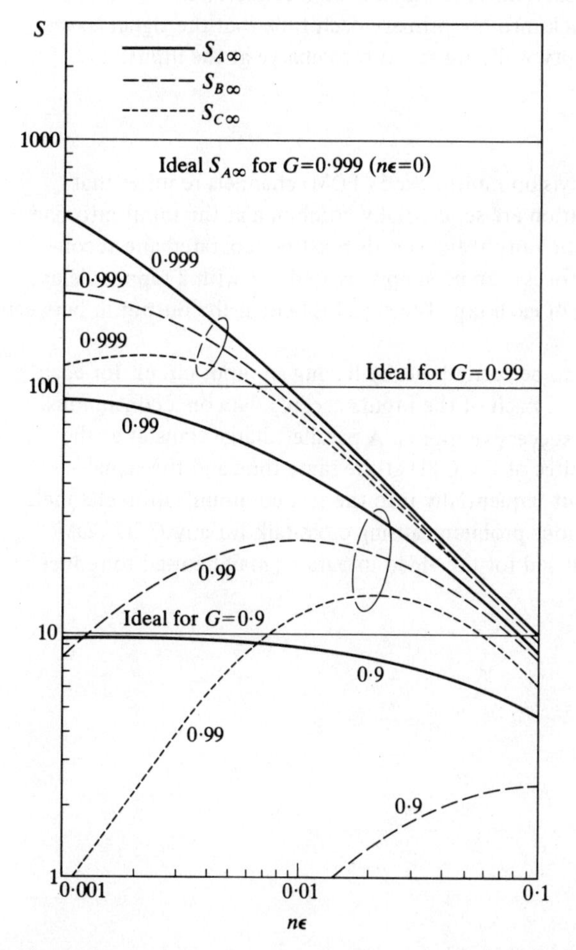

Fig. 6.30 Desired and spurious signal magnitudes in a simple video integrator.

$$S_{A\infty} \rightarrow \frac{A}{1-GA} = \frac{(1-n\epsilon)}{1-G(1-n\epsilon)}$$

$$S_{B\infty} \rightarrow \frac{B}{(1-GA)^2} = \left[\frac{n\epsilon}{1-G(1-n\epsilon)}\right] S_A$$

$$S_{C\infty} \rightarrow \frac{C}{(1-GA)^2} + \frac{GB^2}{(1-GA)^2} = \left\{\frac{(n\epsilon)^2}{2[1-G(1-n\epsilon)]} + \frac{G(n\epsilon)^2(1-n\epsilon)}{[1-G(1-n\epsilon)]^2}\right\} S_A \quad 6.104$$

The magnitudes of $S_{A\infty}$, $S_{B\infty}$, and $S_{C\infty}$ are shown in Fig. 6.30 as a function of $n\epsilon$ for various values of G. For significant integration G has got to be close to unity as discussed in section 6.5. Fig. 6.30 clearly shows the requirement for very small $n\epsilon$ if the spurious signals are to be kept small for useful values of G. These operational difficulties are considerably eased if signals are only allowed to enter the CTD at alternate sampling times as discussed quantitatively by Chowaniec and Hobson (1977). The secondary smeared signal is then collected in the empty well and it can be added back into the primary each time that the signal leaves the CTD. The alternate empty wells are set to zero charge at the input.

6.8 Multiplexing

Communication *via* time division multiplexed (TDM) channels requires that several channels of information are sequentially combined at the input into one high capacity channel. At the output the signals must be separated and reconstructed. This separation process can be simply carried out with a tapped delay line by synchronous gating of each tap. The signal is held at the output in between gating events.

The multiplex process can be carried out by having an input circuit for each tap position in the delay line. Each of the inputs accepts data on a continuous basis and holds it between successive inputs. A parallel charge transfer at the input loads all the storage sites of the CTD at the same time and the signal samples are then clocked out sequentially into the fast communication channel.

Charge smearing is a serious problem causing cross-talk for any CTD TDM system. The measures discussed for the video integrator may be used to reduce its effect.

7

Image Sensing Devices

7.1 Introduction

When a photon is absorbed to produce an electron–hole pair in the depletion region of an MIS capacitor the minority carrier will be stored at the potential minimum. In this way an isolated potential well will contain charge which is proportional to the time integral of the incident optical or infra-red flux. A serial organization of these potential wells in a CCD allows a sequential output charge transfer which is naturally compatible with facsimile communications or raster-scan television. The low-noise performance of CCD charge transfer as described in section 5.4.2 and the possibility of low-noise output sensing described in section 5.3.3 are favourable factors for the use of the CCD as a solid-state camera or image sensor with good low-light-level capabilities. The much poorer noise performance of BB devices generally excludes them from this application even though possible device structures have been considered (Sangster, 1970). A further advantage of the CCD charge-transfer organization compared with previous types of solid-state image sensor (Weimer, 1975) lies in the output sensing where charge is converted to the external circuit signal. The photon generated charge from all sites passes through one output circuit so that problems of non-uniform optical/electrical conversion are much reduced.

Most useful image sensors require about 1000 by 1000 resolution elements. This can be achieved electronically with a two-dimensional array or by a combination of mechanical scanning in one dimension and electronic scanning in the other dimension with a one-dimensional array.

7.2 CCD linear imaging arrays

7.2.1 Optical problems

The simplest linear imaging array has the same construction as the linear CCD described in Chapter 2. One clock phase is held at a high constant potential for charge-containment during integration of the optically generated minority carriers. The other phases are held at low potential. Following the integration period the stored charge is rapidly transferred out of the CCD using the normal clock sequences. This process is repeated for successive line outputs.

If the clock electrodes are metallic they are opaque to the photons which can only enter the CCD depletion regions through the transparent SiO_2 insulator in the gap around the electrodes. This causes a poor light-collection efficiency. Polysilicon electrodes are transparent for optical wavelengths greater than ~450 nm so that they allow the light collection efficiency to be satisfactory except for a poor response in the blue region of the visible spectrum. Other metal oxide materials with better transmission throughout the visible spectrum exist and in principle would be suitable for transparent electrodes (Brown *et al*, 1976) but their employment is not naturally compatible with silicon processing.

If a good response is required throughout the visible spectrum, illumination can be incident on the back surface as illustrated schematically in Fig. 7.1. The

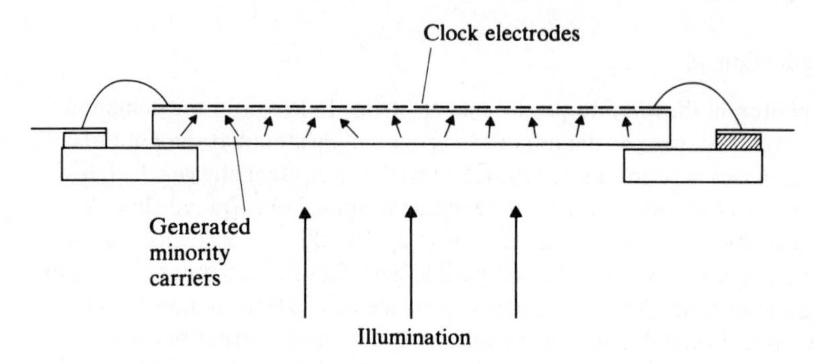

Fig. 7.1 A CCD imager using back-surface illumination.

substrate has to be thinned to a thickness comparable with the electrode dimensions otherwise the random diffusion of charge carriers towards the storage sites can cause image blurring. This requirement usually ensures that the diffusion path is much shorter than the minority carrier recombination length. If infra-red detection is required the substrate must not be made too thin owing to the greater thickness of silicon required for photon absorption, so the problem of image blurring by diffusion remains. Frontal illumination may well provide a better compromise in this case.

7.2.2 Modulation transfer function (MTF)

An important measure of the quality of the conversion of an image to its electrical form is the Modulation Transfer Function (MTF). It is analogous to an electrical circuit transfer function which describes the ratio of output and input signals as a function of frequency. It can be measured by applying an image which has a sinusoidal intensity pattern to the sensor. The electrical output signal amplitude is measured for sinusoidal optical patterns with constant amplitude and various periods (or spatial frequencies). The variation of the electrical signal amplitude with spatial frequency is the MTF. Convergent bar patterns (which are spatial square waves) as illustrated in Fig. 7.2 are often used as the optical object for this measurement.

The simple linear imaging array described above cannot have the ideal unity MTF. During the charge transfer from the CCD, charge smearing will occur and it will generally be worse for higher clock frequencies. In the case of the bar pattern of Fig. 7.2 a cell of high intensity will suffer a first-order smearing into a cell of low intensity when the grating repeat distance has the spatial Nyquist

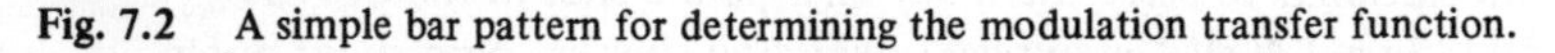

Fig. 7.2 A simple bar pattern for determining the modulation transfer function.

frequency, which is equal to two CCD optical integration cells. If the clock frequency were reduced to give smaller charge smearing there would be image degradation owing to the continuing generation of charge carriers by the incident light. The degradation of MTF by finite charge smearing is the same as that described earlier in equation 4.20 and examples for $n\epsilon$ = 0.1 and 0.2 are illustrated in Fig. 7.3.

The use of buried channel CCD technology would ease the clock rate compromise between charge smearing and extra optical integration effects, providing system constraints did not dictate the clock frequency. The dual

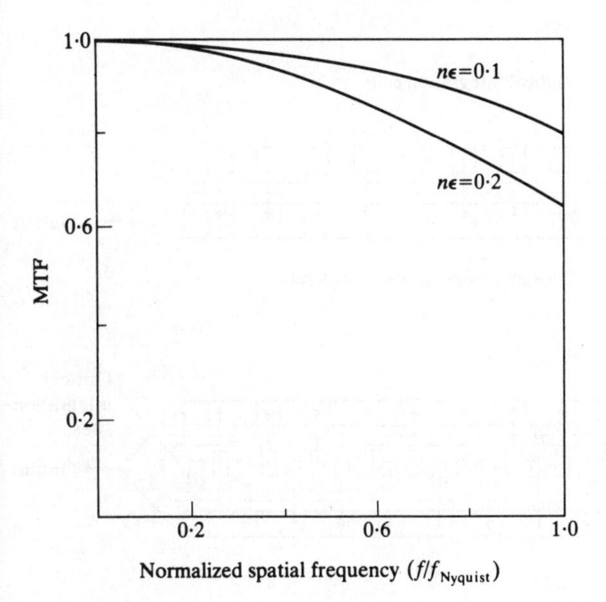

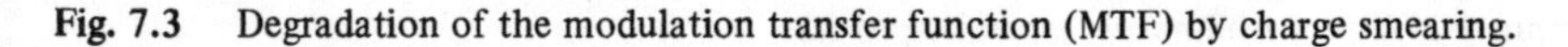

Fig. 7.3 Degradation of the modulation transfer function (MTF) by charge smearing.

advantages, when compared with surface channel CCD, of much faster charge-transfer rate owing to fringing field effects and much lower charge smearing from charge-trapping defects would allow a much higher clock frequency. A further advantage associated with better charge-transfer efficiency in buried channel devices is lowered charge-transfer noise so that this technique is suitable for low-light-level or large dynamic range applications.

Usually the clock frequency is not under the control of the device designer. It is dictated by other system requirements such as compatibility with existing television equipment. Even though a mechanical shutter could be used to blank the optical input during the charge extraction period, it is cumbersome, it has lower reliability than the electronic components and it reduces the optical integration period. A satisfactory compromise between charge transfer and optical integration is to carry them out on separate sites in the CCD as illustrated in Fig. 7.4(a). The optical integration sites are permanent potential wells except that the 'wall' on one side can be removed at the end of the optical integration period by pulsing the transfer gate. The charge is then transferred to a conventional CCD in parallel with the optical integration sites and sequentially transferred through the output. An opaque covering must be put over the CCD transfer structure to avoid any spurious optical integration during transfer to the output. As was seen in Fig. 7.3 the MTF is degraded by finite $n\epsilon$.

Since typically 1000 elements may be required for one line of optical information, stringent requirements are placed on low charge-transfer inefficiency during the output clocking period. Twofold improvement may be achieved in the bilinear arrangement illustrated in Fig. 7.4(b). Practical details have been given by Kim and Dyck (1973) and Tompsett *et al* (1973).

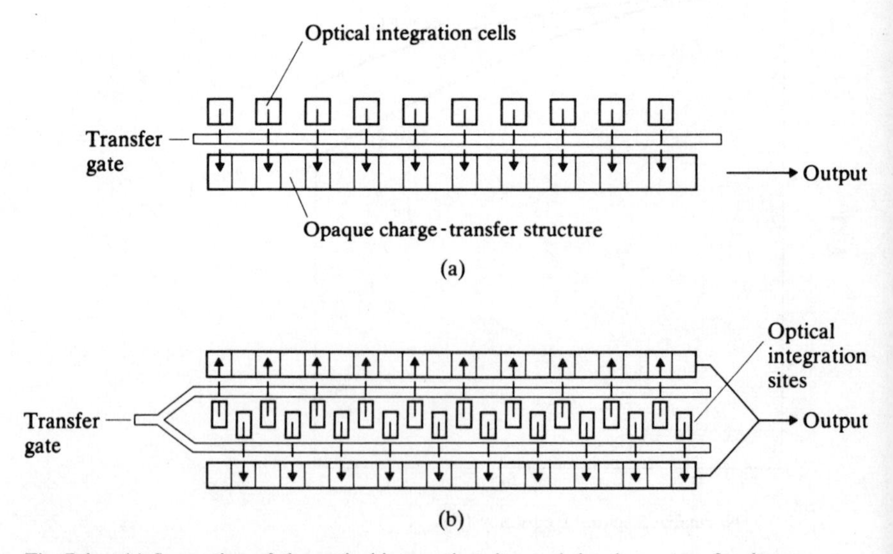

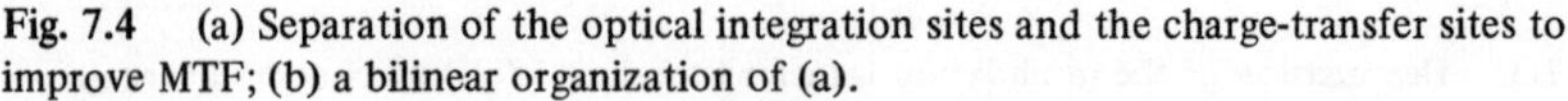
Fig. 7.4 (a) Separation of the optical integration sites and the charge-transfer sites to improve MTF; (b) a bilinear organization of (a).

The MTF of back-surface illuminated sensors is degraded by random carrier diffusion effects from the back to the front surface. Alternatively, the lateral spreading of charge will limit the spacing of charge stores on the front surface so that there is an implied limit to the resolvable spatial frequency. This limit has been calculated for other types of solid-state imager by Crowell and Labuda (1969) and some examples are illustrated in Fig. 7.5.

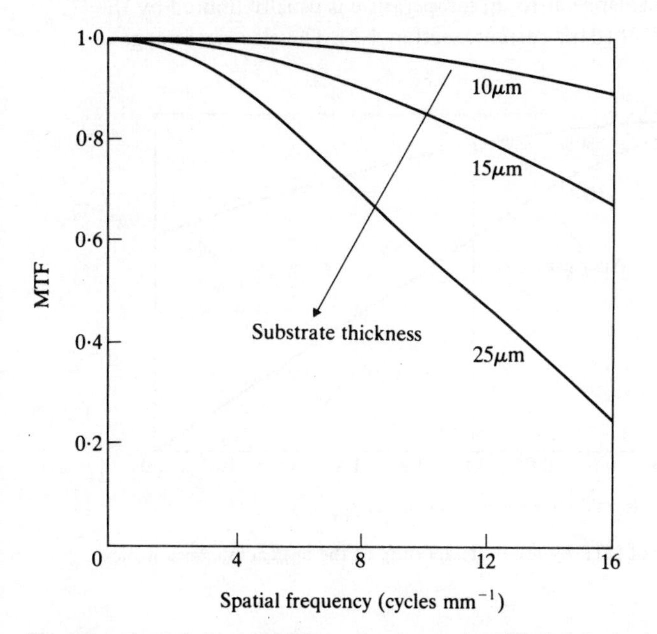

Fig. 7.5 Degradation of MTF by minority carrier diffusion for back-surface optical reception.

A further source of degradation of the MTF arises from the finite size of the optical integration sites. A unity MTF at all spatial frequencies could only be obtained if the sensors had zero width in which case the array would have low sensitivity. If the optical integration site has a finite uniform aperture, a $(\sin x)/x$ type of response is superimposed on the overall MTF in a way which is analogous to the bandwidth limits of the input circuit of signal processing CCD. This was described in section 5.6. In the limit when the aperture of each optical integration site is equal to the spacing of adjacent sites the $(\sin x)/x$ pattern has its first zero at twice the Nyquist frequency (i.e. MTF $= 0$) and the MTF has fallen to 0.64 at the Nyquist frequency. In later discussions of two-dimensional arrays the MTF for signals perpendicular to a line will also be important. For the separate site organization illustrated in Fig. 7.4(a) the optical integration sites may occupy 50% of the available dimension. In this case the MTF is only degraded to ~0.9 at the Nyquist frequency. These modifications to the MTF are illustrated in

Fig. 7.6. In some cases it may be desirable to use this degradation of the MTF as an anti-alias feature (Campana and Barbe, 1974). This is perhaps an important feature for broadcast television where picture quality is of great importance. In military applications the information control is likely to be of greater interest so some aliasing of information may be allowable.

7.2.3 Limitations at low light levels

Low-light-level performance at room temperature is usually limited by the residual leakage current (dark current, section 4.3). The electrical image

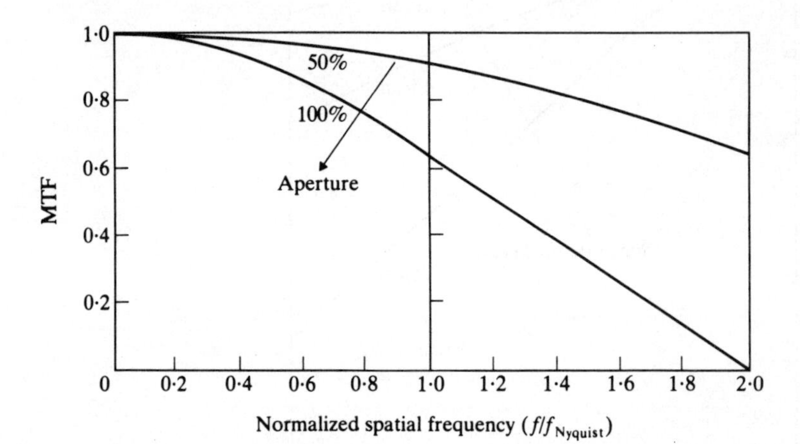

Fig. 7.6 **Degradation of MTF by the finite aperture of the optical integration sites.**

contrast is reduced owing to the mean level of added charge from the dark current. More seriously, the superimposition of fixed pattern noise on the desired image owing to the spatial dependence of the dark current causes loss of information. At room temperature, assuming an optical integration site of 20 μm by 20 μm, a mean dark current of 10^{-5} A m^{-2}, and an integration time of 1/25 s (corresponding to broadcast television) the spurious electrical output per pixel (i.e. from each site) would be 1000 electrons. This is significantly higher than the charge-transfer noise and output noise described in section 5.4.2. If better dark current performance is required the CCD must be cooled because the dark current is expected to decrease exponentially with the inverse of the absolute temperature as illustrated in Fig. 7.7 (Dyck and Jack, 1974). Temperatures of −30°C to −50°C should then ensure that the noise performance is the limiting factor. The noise performance of the output sensor will be particularly important (section 5.4.2) and techniques such as the distributed floating-gate amplifier (section 5.3.5) and correlated double sampling (section 5.4.3) must be used. When low-noise performance becomes important it is essential to use a buried channel CCD. The necessity to introduce a fat zero for surface channel devices in order to have acceptable charge-transfer efficiency also causes the

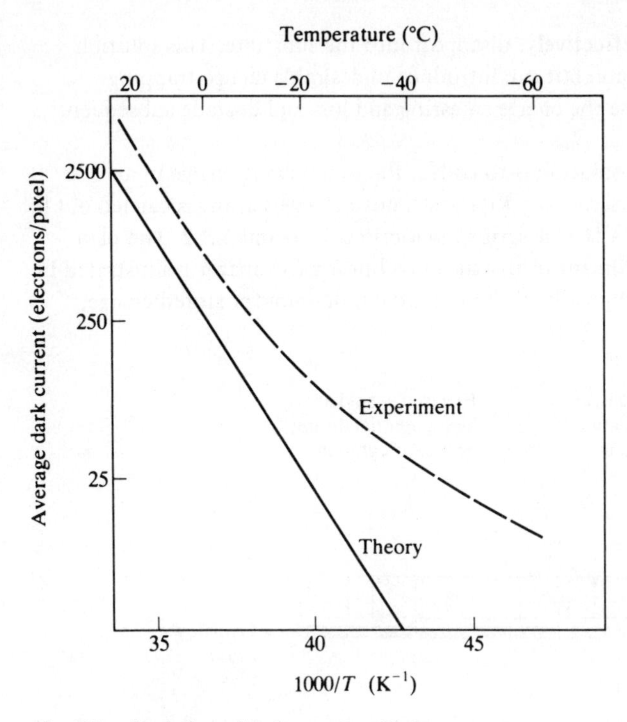

Fig. 7.7 Variation of dark current with temperature.

introduction of dominant Johnson noise at its injection point (sections 5.4.2(ii) and 5.4.2(iv)) as well as noise caused by the trapping and release of electrons. The broader depletion layer for a buried channel device when compared with a surface channel device should theoretically cause a larger dark current in the former device but this fact is not usually of practical value because of the other dominant defects in the surface channel device.

One undesirable consequence of the low dark current is the emptiness of residual charge-trapping states in buried channel CCD. In effect the dark current acts as a kind of fat zero. At low temperatures the extra empty charge-traps may cause an enhanced charge smearing. Its practical effect has been described by Wen (1975).

7.2.4 Saturation limits and blooming control

Saturation effects occur when the amount of integrated charge fills the collecting well so that its surface potential becomes high enough to allow charge spillage into neighbouring wells. This effect causes the images of bright objects to 'bloom' by spreading and appearing as a bright 'pool' of light when they are subsequently displayed. In the simple linear CCD imager the cells with low bias (during the integration period) that are used to separate the integrating cells may be biased into accumulation so that overflowing charge can recombine with majority carriers and, effectively, disappear into the substrate. This solution limits the blooming region but can introduce undesirable charge-trapping problems which increase the charge smearing and loss and degrade subsequent image information.

A more satisfactory solution is to collect the overflowing charge in an 'overflow drain', in the same way that destructive charge sensing is carried out in the output diode of a CCD as described in sections 2.16 and 5.3.2. The plan view of a simple embodiment of this idea in a linear CCD imager is illustrated in Fig. 7.8. The associated profile of the potential minimum for stored charge,

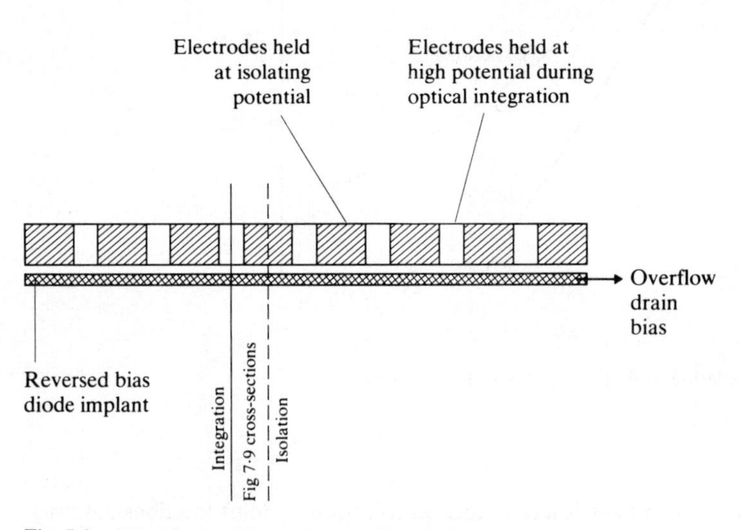

Fig. 7.8 The plan position of an overflow drain.

perpendicular to the transfer direction is illustrated in Fig. 7.9(a). The potential energy profile, shown by the solid line, corresponds to an empty integrating well. The potential barrier between this well and the overflow drain is necessary to avoid overflowing charge spilling into non-saturated potential wells at other points of the array. There are several ways of implementing the potential barrier using techniques already discussed in Chapter 2. In Fig. 7.9(b) and 7.9(c) the techniques are the same as those used for two-phase clocking systems for raising the surface potential with a p^+ implant in a p-type substrate or use of a thick oxide respectively. The technique used in Fig. 7.9(d) is analogous to the use of the output gate to control the flow of charge to the output diode of a CCD as discussed in section 5.3.2. It has the advantage that it gives control of the potential barrier height between the integrating well and the overflow drain. The implementations illustrated in Fig. 7.9, appear technologically cumbersome but are simpler to integrate in area imaging devices, particularly of the frame-transfer type. The potential profile shown with the broken line in Fig. 7.9(a) is that in

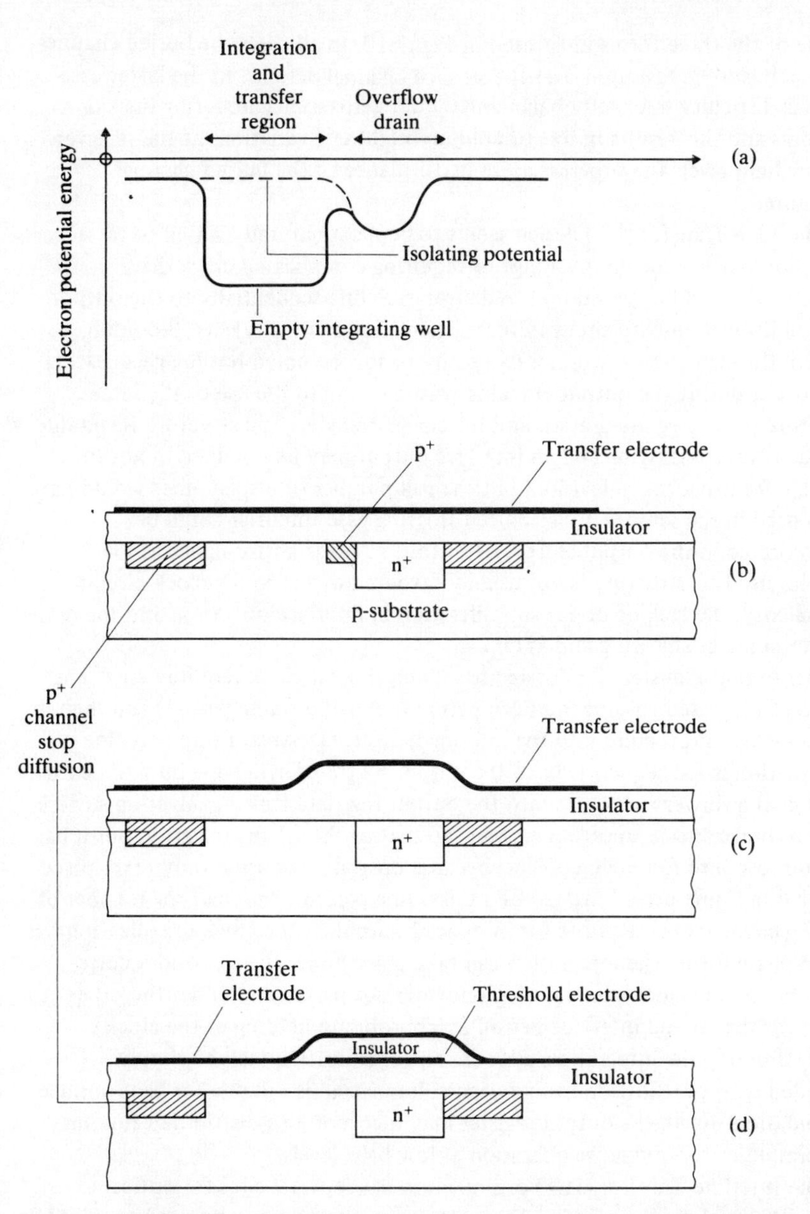

Fig. 7.9 Cross-sections through the charge storage region and the overflow drain of Fig. 7.8. (a) potential energy profiles; (b), (c) and (d) are various technological realizations of (a)

the cross-section through the overflow drain and a neighbouring transfer electrode held at isolation potential.

7.3 CCD area imaging arrays

The above design considerations may be incorporated in an area imaging device in one of the three forms illustrated in Fig. 7.10. In all cases the buried channel approach is to be favoured over the surface channel designs. In the latter case there is difficulty in exactly balancing the fat zeros in neighbouring lines or columns and this results in line or column brightness variations at the receiver. At low light levels the superior noise performance of the buried channel device is required.

The Line Transfer (LT) design seems to be most naturally suited to raster scan television but has the disadvantage of requiring complicated clock drive circuitry (line address circuitry) to deliver each line sequentially to the output. This additional on-chip circuitry increases manufacturing rejects. Blooming control through overflow drains can easily be incorporated but image smearing may occur during the output clocking period owing to the use of the same structure for charge integration and for charge transfer. The structure is suitable for back-surface imaging and an interlace output may be obtained in a form suitable for broadcast television, but the full number of display lines would have to be used in the sensor. A distributed floating-gate amplifier could be incorporated in the output shift register for low-noise performance at low light levels. This structure is not usually favoured owing to the clock circuit complexity, the lack of device simplification in interlace operation and the non-optimum image smearing and MTF.

The Frame Transfer (FT) design has much simpler clock requirements. The image is integrated for one interlace period (i.e. half a frame period) and then it is all clocked out rapidly into the intermediate storage area. From here line information is fed sequentially to the output by transferring one bit from each column at a slower clock rate into the output register. This organization suffers from a similar image smearing problem to that of the LT organization but it has a faster response to moving objects because optical integration only takes place over half a frame period instead of a full frame period. Only half the number of sensing elements are required for interlaced outputs if the clock organization has a two-phase form. The integration can take place under the electrodes corresponding to one clock phase during one interlace period and under the other phase for the second interlace period by appropriate holding of the clock voltages during the integration periods. The organization may be simply provided with overflow drains between columns and is suitable for back-surface illumination. Again the output register may incorporate a distributed floating-gate amplifier for low-noise operation at low light levels.

The Interline Transfer (ILT) organization has separate sites for optical integration and charge transfer. This gives it good image smearing and good MTF performance but the opaque charge-transfer regions reduce its light-collection efficiency and make it unsuitable for back-surface illumination. Interlacing can be carried out with a full frame number of optical integration sites from only a half frame number of charge-transfer sites. The charge transfer is carried out from imaging columns to storage columns on opposite phases of a two-phase clock for successive interlace periods. Overflow drains can be incorporated between columns but their technological layout and that of the two types of charge handling region is somewhat complex as may be ascertained by reference

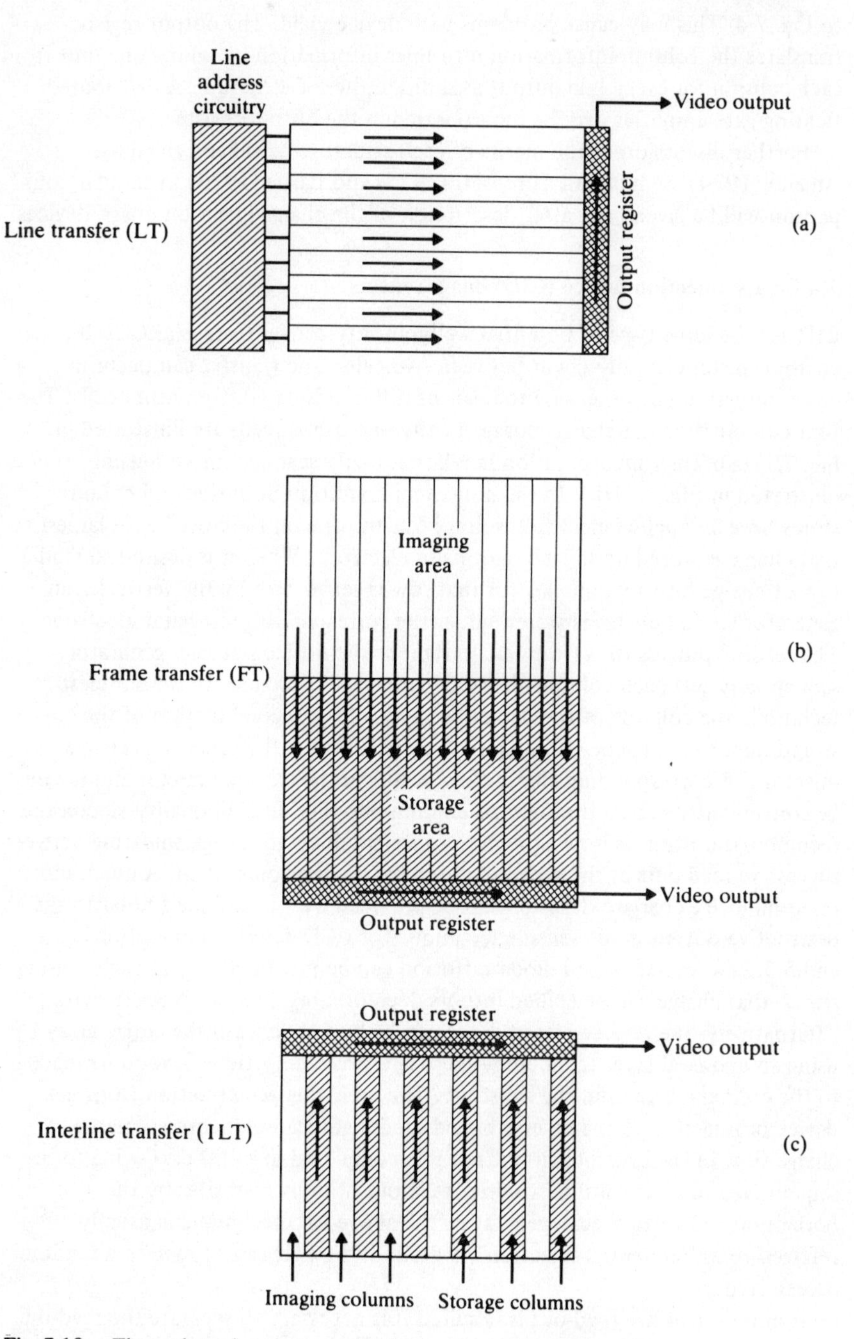

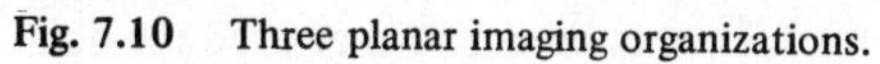

Fig. 7.10 Three planar imaging organizations.

to Fig. 7.4. This may cause problems with device yield. The output register translates the column information into lines information by taking one unit from each column for every line output as it did in the FT structure. A distributed floating-gate amplifier may be incorporated in the output register.

Further discussion of the merits of each structure have been given by Amelio (1974); Sequin and Tompsett (1975) and Barbe (1976). A tabular comparison will be given later after description of the charge-injection image devices.

7.4 Charge-injection device (CID) image sensors

CID use the same type of potential well minority carrier storage as CCD but charge transfer can only occur between two cells. The transfer can occur in either direction but otherwise the pair of cells are isolated from other cells. The four combinations of charge storage in any one pair of cells are illustrated in Fig. 7.11(a). Their incorporation in a horizontally scanned image sensing array is illustrated in Fig. 7.11(b). In the non-select condition both row and column stores have an applied electrode voltage but the one on the row, V_R, is largest so that charge is stored under the right-hand electrode. When it is desired to read a row of charge information, V_R for that row is set to zero by the vertical scan generator so that charge is stored under the corresponding left-hand electrode. The reading process runs from left to right as the horizontal scan generator sequentially sets each column electrode voltage, V_c, to zero. In the simplest technique the collapse of the depletion layer allows recombination of the stored minority carriers with majority carriers. This is the process of charge injection. A corresponding charge flow occurs in the external circuit and it can be conveniently sensed through the column bias line. In high quality silicon the recombination time is long enough to cause trouble with image smearing between successive read-outs at the speed required for a television system. A much more rapid minority charge extraction can be provided by a technique similar to the destructive output diode sensing technique for CCD described in sections 2.16 and 5.3.2. A reverse biased diode diffusion can be provided around each sensing site so that charge can be spilled into its depletion region as V_c is set to zero. Alternatively, the reverse biased diode can be buried beneath the entire array by using an epitaxial layer technology. Separate bias connections have to be made to the epitaxial layer and the substrate. The rapid charge extraction from the device or injection to the reverse biased diodes still causes a corresponding charge flow in the external circuit but it is completed in ~100 ns. Owing to the sequential addressing and destructive read-out of individual sites by the horizontal and vertical scan generators the above CID technique is usually referred to as 'sequential injection'. If desired, the CID can be used in a random access mode.

If non-destructive read-out is required it is necessary to separate the read-out process from the injection process. The technique is similar to that used for non-destructive read-out of CCD information as described in section 5.3.3. The cell pairs are all initially held in the non-select condition shown in Fig. 7.11(a) and the horizontal scan generator is set to isolate the column line electrodes with the

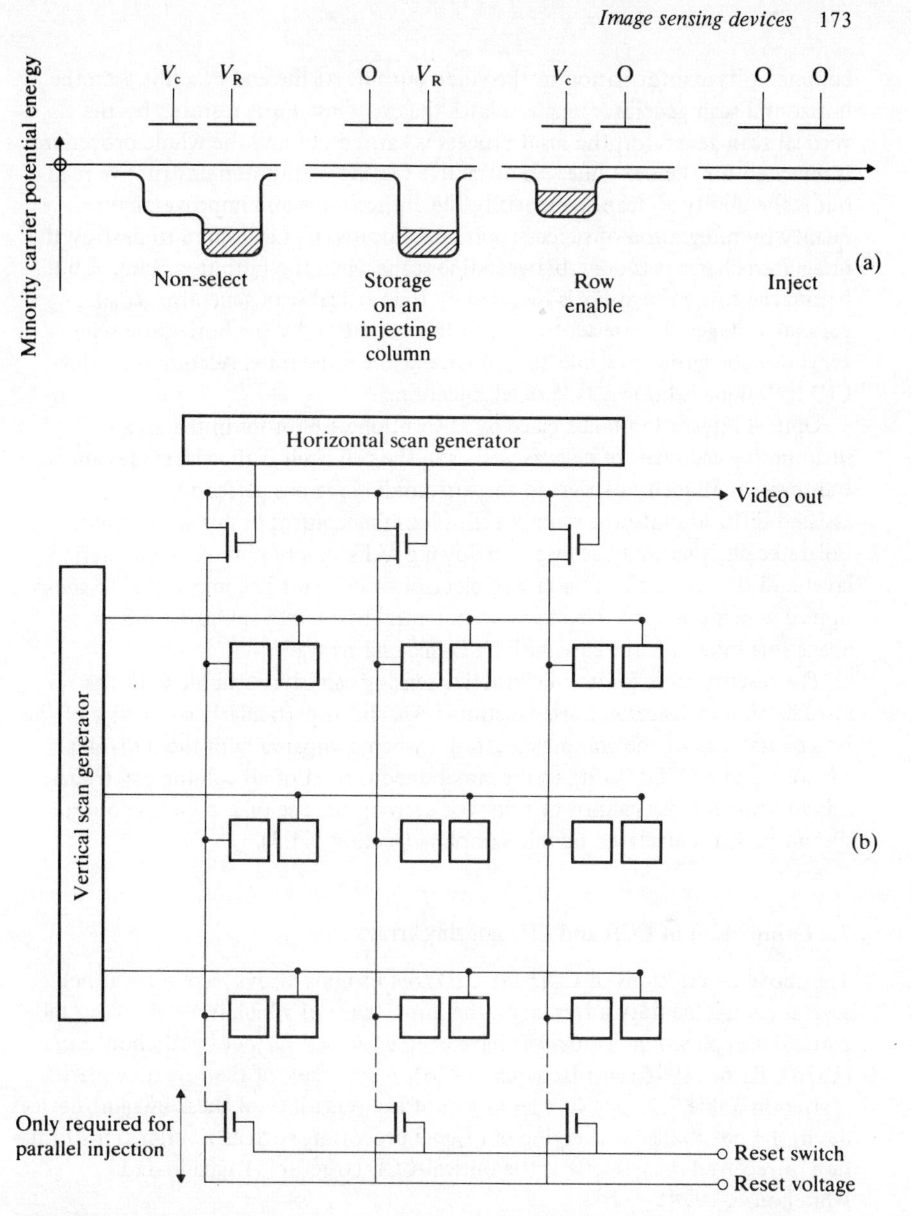

Fig. 7.11 (a) Potential wells in a charge-injection device (CID); (b) plan view of the organization of a CID imaging array.

FET switches. Each column electrode is then reset to a predetermined voltage by the lower FET reset switches and again isolated. Line reading is carried out by setting V_R on the desired line to zero so that charge is transferred under the column electrodes in the row enable form. This causes the column electrodes to change potential as dictated by their capacity and the amount of transferred charge. The horizontal scan generator is then used to transfer sequentially the column voltage information to the video output. At the end of a line scan the horizontal scan generator again isolates the columns, V_R is restored by the vertical scan generator, the reset process is carried out and the whole procedure is repeated for the next line. An attractive feature of this non-destructive read-out is the ability to scan sequentially the entire frame and improve picture quality by integration of successive frame outputs. If it is desired to destroy the integrated charge it is done between line scans when the output is blanked but before the row voltage V_R is restored by the vertical scan generator. If all column voltages, V_c, are set to zero in this condition by the horizontal scan generator the entire row injects its charge at the same time. Accordingly, this CID technique is known as Parallel Injection.

Optical integration takes place by absorption of photons in the area surrounding each pair of cells as well as in the cell itself if the electrodes are transparent. Minority carriers in the surrounding regions suffer a field assisted diffusion into the storage cell. Blooming control in this structure of isolated cells is natural because overflowing wells simply remove the depletion layer and the recombination rate of electron–hole pairs becomes equal to their optical generation rate. There are no potential barriers to spill over and cause image smearing as is the case with CCD imaging arrays.

The reset process for non-destructive sensing can cause trouble with the introduction of Johnson noise (section 5.4). This is particularly so in view of the larger total area of the column electrodes when compared with the individual electrodes of a CCD. Owing to the simultaneous reset of all column electrodes, this additional noise cannot be removed even by the use of a low-noise output system such as correlated double sampling (section 5.4.3).

7.5 Comparison of CCD and CID imaging arrays

The above descriptions of CCD and CID area imaging arrays have shown their several complementary advantages, the importance of which is weighted by the particular application. Trade-offs have been discussed further by Michon *et al* (1975), Barbe (1976) and Baertsch (1976). A summary of their relative merits is given in Table 7.1. One further aspect of the versatility of these imaging devices lies in the possibility of carrying out signal processing such as Fourier transformation of received images within the optical/CCD structure (Lagnado and Whitehouse, 1974).

7.6 Infra-red CCD and CID devices

The problems of vision in the dark and imaging of objects from their own thermal radiation may have solutions in an infra-red television camera using the foregoing techniques. The atmospheric 'windows' where water vapour and carbon dioxide molecules do not cause significant absorption occur for free space wavelengths of ~1 μm, 2 μm to 2.5 μm, 3.5 μm to 4.2 μm and 8 μm to 14 μm. Unfortunately, the silicon used in CCD and CID devices is largely transparent to the three longer

	Frame Transfer CCD	*Interline Transfer CCD*	*CID*
Versatility	Front- or back-surface illumination. Can be used for TDI.	Front-surface illumination only. Separate sensors make video signal processing possible.	Front-surface illumination only with epi-CID. More complicated injection sites for back surface illumination. Random access readout possible.
Spectral sensitivity	Large fraction of theoretical silicon response for back-surface illumination. Front-surface layers cause optical interference modifications to response.	Shielded front-surface and complicated deposition patterns reduce sensitivity by more than twofold.	Good front-surface collection efficiency in low density arrays. Large fraction of silicon response in back-surface illumination.
Use of silicon chip area	<50% available for imaging	<50% available for imaging	~90% available for imaging
Anti-blooming control	Accumulation around cells or overflow drains give good blooming performance.	Overflow drains can be placed between columns giving good blooming performance.	Accumulation around cells provides excellent anti-blooming control.
Low-light and noise performance	Cooled, buried channel devices and low capacity collection nodes give good noise performance and good low-light performance.	Similar comment to frame transfer CCD.	Higher capacity charge collection nodes and reset noise give a problem which is offset by the ability to integrate over many frame cycles with the non-destructive readout.
MTF at spatial Nyquist frequency			
Vertical with interlace	0 (overlapping pixels on interlace)	~0.6	0.6–0.9
Horizontal	~0.6	~0.9	0.6–0.9
Infra-red performance	Suitable for Si-based charge transfer	Suitable for Si-based charge transfer	Suitable for compound semiconductors owing to independence of charge-transfer efficiency
Special problems	High-speed vertical transfer	Complex cell	Fixed pattern noise

wavelength windows because the corresponding photon energies are less than the bandgap energy.

There are four basic approaches to the problem.

(i) The charge read-out structure can be fabricated on a semiconductor with suitable spectral response such as InSb, HgCdTe, PbSnTe, PbS etc. Al_2O_3 insulators or anodic oxides of the semiconductor may be used for the insulator region of the MIS structure. The small bandgaps and poor quality of these materials cause considerable problems with poor charge-transfer efficiency and high dark current, and operation has usually to be carried out at an ambient temperature of 77 K or less.

(ii) Hybrid imaging systems can be made by using a suitable material to detect radiation then passing the photoelectrons into a silicon CCD or CID using diode and gate input circuits similar to those of the signal processing CCD. This approach can remove the problem of poor charge-transfer efficiency in (i) but has noise and interface problems associated with the provision of individual inputs at each sensing point of the structure

(iii) In order to utilize the good quality silicon MOS technology, suitable dopants can be introduced to allow generation of electrons alone from localized energy levels in the bandgap of silicon. Gallium is one suitable impurity and can provide a useful detection sensitivity in the wavelength range 3 μm to 17 μm. Introduction of impurities can cause trapping difficulties unless great technological care is taken.

(iv) Another approach to utilize the advantages of silicon MOS technology is to generate the stored charge in the CCD or CID from photon absorption in an adjacent Schottky barrier diode. A suitable metal must be chosen for the desirable wavelength range so that electrons can be excited over the barrier from the metal to the silicon depletion region.

More detailed discussions of these various approaches have been given by Elliot (1976) and in General References. Whichever approach is taken, the objects to be viewed usually have a low contrast of 1% or less so that dark current and

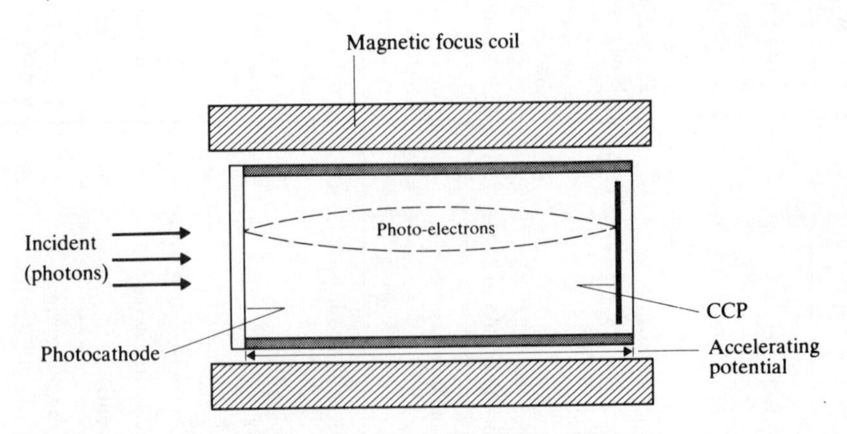

Fig. 7.12 An electron beam imaging device with a CCD or CID target.

other detection non-uniformity is a considerable problem. One way to enhance the image is to use the time delay and integrate technique (TDI). In this, the infra-red camera is mechanically scanned so that successive frame outputs from the same point of the object come from adjacent detection/store cells. In this way the fixed pattern non-uniformities from many units of the imaging area are averaged in each section of the integrated image. The integration can be carried out either in the image sensing CCD in which the information is transferred cell by cell in step with the scanning motion or it can be carried out in a subsidiary CCD store which collects each frame output with the appropriate displacement between frames. This procedure gives a $\sqrt{N}$ improvement of signal/background noise where N is the number of frames integrated and the background noise is that associated with the object area. A similar technique can be used in the visible spectrum for observation of dim objects such as satellites or astronomical objects.

7.7 Electron beam CCD and CID image sensors

In the electron bombarded semiconductor (EBS) mode of operation the CCD or CID receives electrons rather than photons in a back-surface illumination arrangement (Holeman and Gardner, 1976). The optical image is converted to electrons by a photocathode in a vacuum tube (Fig. 7.12). The emitted electrons are accelerated as in a cathode ray tube and focused on to the CCD or CID. When they enter the semiconductor back surface they are absorbed and each one produces a large number of electrons (typically greater than 1000) owing to the ionization arising from their high energy. These secondary electrons are collected by the appropriate potential wells and transferred to the output in the usual CCD or CID manner. It is not clear that the large electron multiplication will produce a better low light viewing device than direct utilization of a buried channel CCD owing to problems of focusing accuracy (and consequently image smearing) and noise in the electron bombardment.

8

CCD Digital Memories and Logic Functions

8.1 CCD memory devices

The high density of information storage sites in a CCD are attractive for computer storage of binary digits. The simple electrode structure for each bit of information suggests that their surface storage density should be larger than that of MOS Random Access Memories (RAM) (Terman and Heller, 1976). The yield of working silicon chips for any of these technologies decreases at more than a linear rate with increasing area. The smaller area requirement and simple organization of a CCD results in a smaller cost per storage bit than for a MOS RAM. The serial nature of CCDs allows much less frequent access to the stored information but for computer applications requiring transfer of large blocks of information their clock rate can be faster than the information transfer rate of MOS RAMs. The serial transfer in CCD causes averaging of the dark current signal whereas MOS RAM must allow for worst case dark currents in an entire array.

The high information transfer rate of CCD also makes them attractive when compared with magnetic disc stores. The access time can be 100 to 1000 times shorter than that of discs but the storage capabilities are smaller and the cost per bit of storage is higher.

Charge-transfer inefficiency and generation of extra charge from dark currents all conspire to corrupt the binary information and limit the maximum sequential length of a store. The former effect in surface channel CCD can be worse than in other applications of CCD owing to the requirement for minimum surface area. This causes the 'sides' of a transfer channel to occupy a large fraction of the total channel width. The charge smearing which is caused by the surface states in these channel sides cannot be reduced by a fat zero (section 4.2.2).

The consequence of all these features are usually summarized in the form of access time *versus* cost per bit or capacity diagrams as shown in Fig. 8.1. Any decisions on the value of CCD memories are much more dependent on the particular application than is suggested by these simple considerations. Before more detailed factors can be considered it is necessary to describe four memory organizations possible with CCD.

The serpentine structure shown in Fig. 8.2 uses regenerators for the binary information after charge smearing and dark current has caused signal degradation. The regenerators also turn the bit stream around so that it will fit on to the silicon chip. In essence the regenerator contains a destructive read-out stage (section 5.3.2) followed by a differential amplifier, which determines whether the output is greater or less than the mean output signal and a charge input

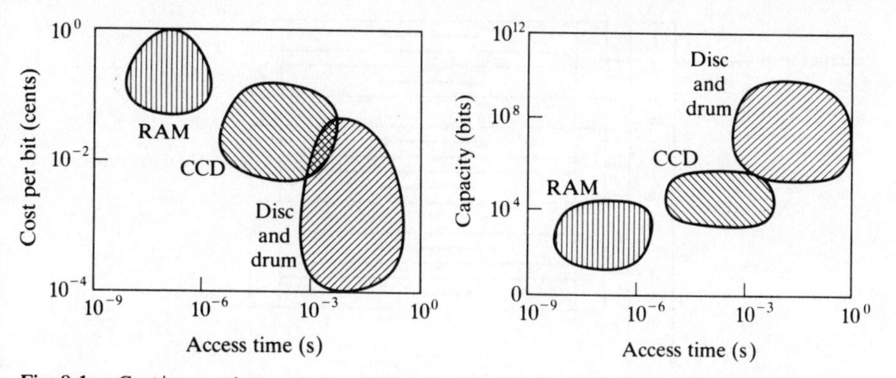

Fig. 8.1 Cost/access-time and capacity/access-time relationships for some computer memories.

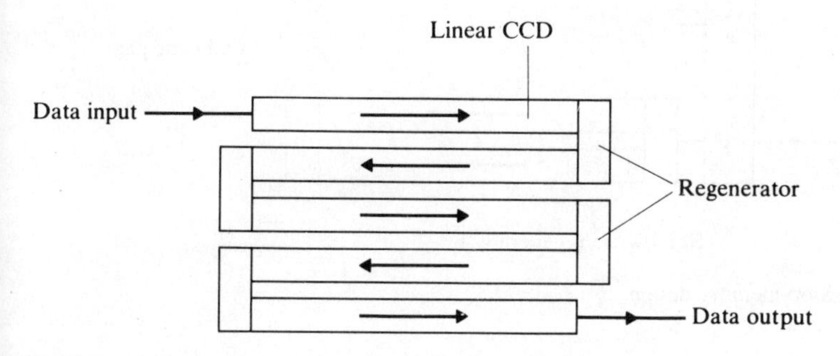

Fig. 8.2 A serpentine memory design.

circuit for the next CCD line. Typically, there may be 128 or 256 charge stores in each CCD line. The differential decision in the regenerator requires a half-charge generator which must be accurate, relatively, as a direct consequence of the processing technique so that variations from chip to chip and line to line are automatically taken into account. The serpentine structure has a very long access time and large storage capacity.

The loop organization has similar layout to the serpentine organization except that the information is recirculated in pairs of CCD lines as shown in Fig. 8.3. It has been referred to as the 'drum' organization because of its organizational similarity to the magnetic drum memory. Each pair of CCD register lines has a regenerator and turn around circuit at each end. The left hand regenerators also provide for destructive or non-destructive reading and writing (Chou, 1976). Access time in this system is set by the transfer time through one CCD line pair and so is faster than the serpentine array. Further speed increase can be obtained by storing the individual bits of an entire binary word in corresponding time slots in each recirculating line pair by appropriate design of the line address selector. Both the serpentine and loop organizations are referred to as synchronous stores because the shift of all bits occurs simultaneously.

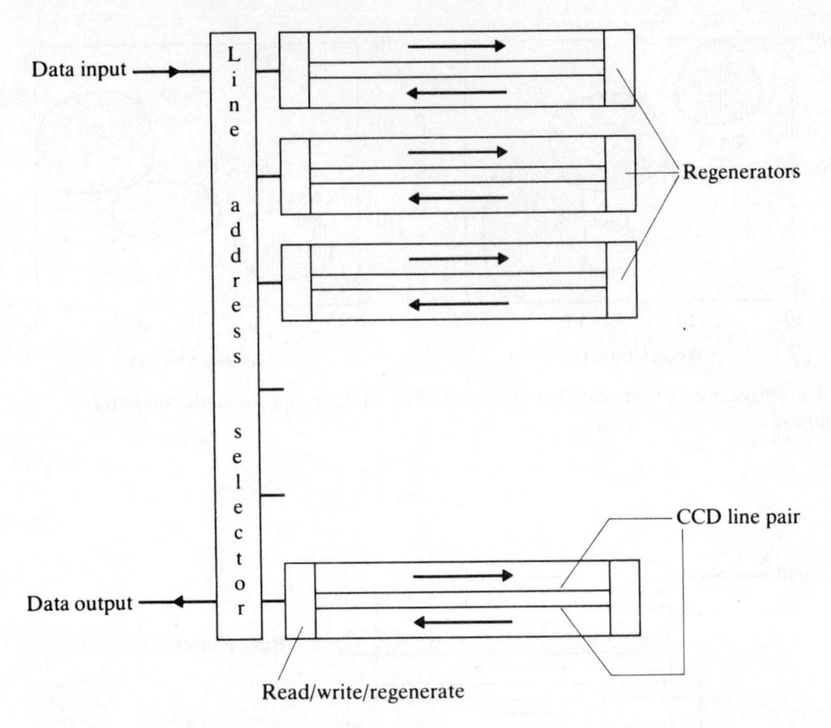

Fig. 8.3 A loop memory design.

A neat technique to reduce drastically the number of charge-transfer events in a long serial store is found in the series–parallel–series (SPS) stores of Fig. 8.4 (section 5.7). The horizontal CCD registers at input and output work at high clock frequency. Each time that the input line is filled it is entirely transferred

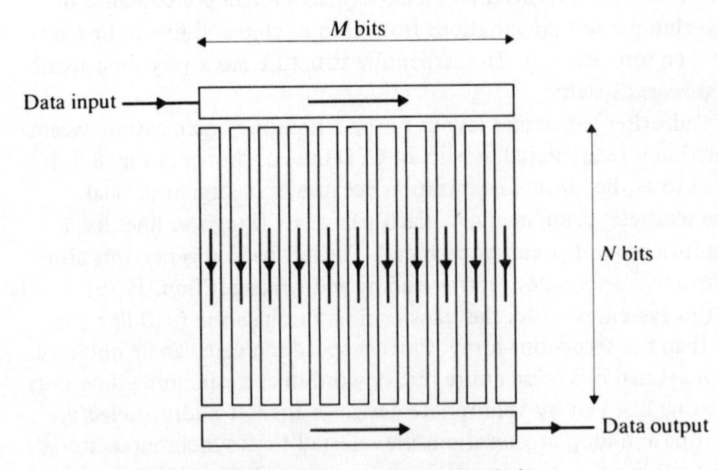

Fig. 8.4 A Series–Parallel–Series (SPS) memory design.

into the CCD columns and one row at the bottom of the columns is transferred into the output line. Only $(M+N)$ charge-transfer cycles occur for $(M \times N)$ storage elements. There is the further advantage that the transfer process in the columns is M times slower than in the input and output rows, so that the overall bit rate can be high without incurring serious charge-transfer penalties. The SPS arrangement has a convenient geometry for non-overlapping layout of its clock lines and has relatively simple peripheral circuitry.

The line addressable random access memory (LARAM) shown in Fig. 8.5 was designed to reduce the access time in a CCD store. Each line is held with stationary clock, after reading in, except for occasional regenerative circulations to counteract the dark current effects. When one line of information is required, a fast clock sequence is applied under the control of the line address selector and the data is read out. This random access can only apply to an entire line of information but it does not have to wait until other lines of information have gone through a prescribed sequence. An additional advantage of this organization is the low clock power requirement as a result of the non-continuous charge-transfer process. The SPS and LARAM organization are non-synchronous owing to the different transfer rates in different parts of the store.

8.2 Memory design considerations

8.2.1 Packing density and clock design

When large quantities of data have to be stored, the area per bit must be as small as possible both to minimize the clock power dissipation (by reducing capacity

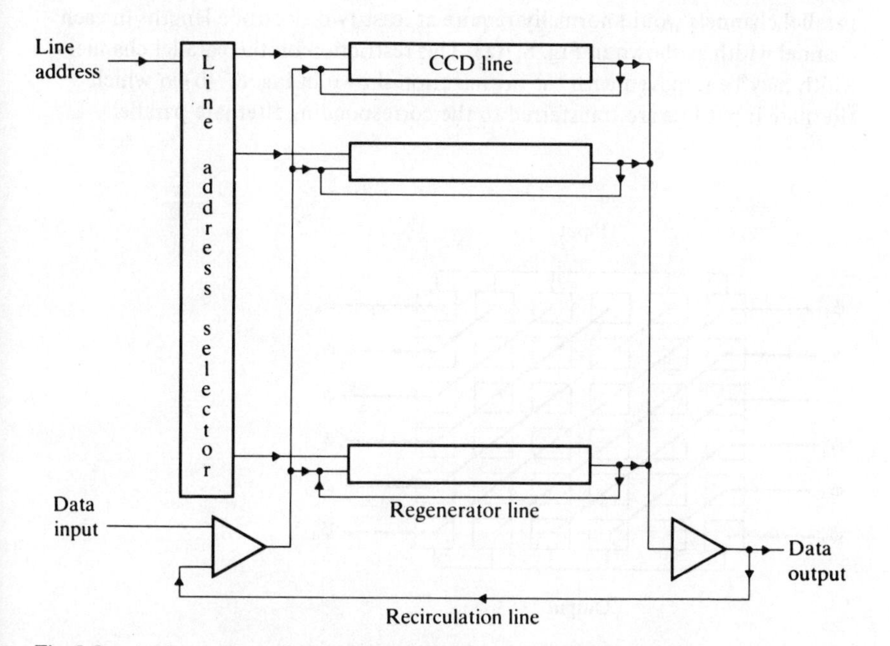

Fig. 8.5 A Line Addressable Random Access Memory (LARAM) design.

per bit) and to optimize the yield from the production process which decreases rapidly with increase of chip area for a given layout complexity.

Storage requirements for large serial blocks of data where random access is less important than high data rate and high packing density are naturally suited to the SPS organization. Several detailed memory designs of this type have been described by Rosenbaum *et al* (1976), Tchon *et al* (1976) and Mohsen *et al* (1976). It has the least amount of peripheral circuitry, owing to the simple single input and output interfaces, and the information packing density can be increased by use of the electrode/bit clocking sequence that was described in section 2.5.4 and Fig. 2.17. Each electrode has the built-in directionality of the two-phase electrode so each one occupies two units of the minimum area allowed by the photolithographic production rules and the CCD channel width. If the electrode/bit clocking sequence was used with N phases, then N of the electrodes would contain N-1 bits of data. This would give an area 'efficiency' of $2N/(N-1)$ minimum area units per bit compared with three minimum area units for simple three-phase clocking. Large N requires complicated clock driving circuitry which uses extra chip area so that the diminishing rewards of increasing area efficiency at high N often cause a compromise to be struck with $N < 8$. A natural evolution of the SPS structure using electrode/bit clocking is illustrated in Fig. 8.6. With its diagonally synchronized clock electrodes this organization sequentially self-samples the input horizontally along the top of the columns. It does not require a separate input register nor does it require an output register.

The minimum area consideration can only be met if the parallel channels have their minimum width. Transfer from the serial input register to the parallel channels would normally require at least two electrode lengths in each channel width as shown in Fig. 8.7(a). This restriction on the parallel channel width may be removed with the organization shown in Fig. 8.7(b) in which alternate input bits are transferred to the corresponding alternate parallel

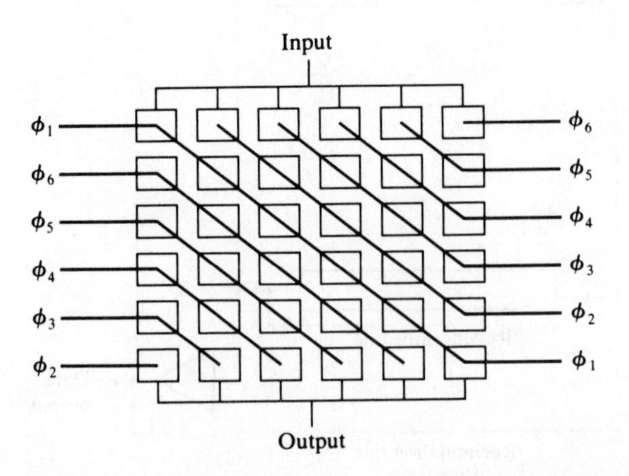

Fig. 8.6 A memory array design using electrode/bit clocking.

channels during one phase of the input register's two-phase clock sequence. The remaining alternate input bits are transferred to their corresponding parallel channels during the next phase of the input register's clock sequence.

The maximum block size that can be accommodated in one SPS register is limited by yield considerations, charge smearing and dark current as illustrated in the following example. The electrode area used in SPS block storage is as small as possible and 15 μm by 30 μm may be taken as a representative size. The

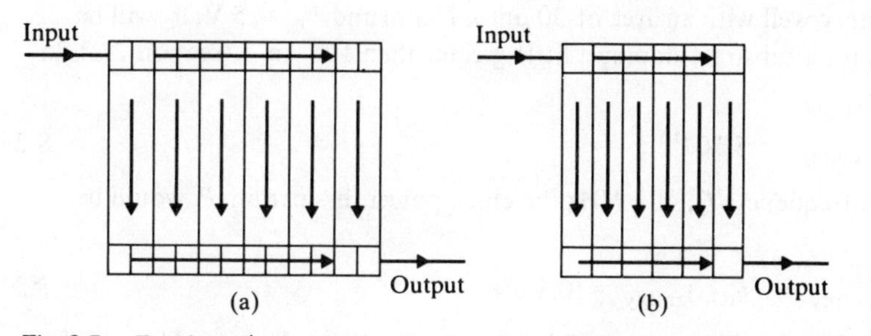

Fig. 8.7 Two input/output register designs for an SPS memory. The design in (b) allows higher packing density.

charge-transfer efficiency is degraded by such narrow channels (section 4.2.2) and the dark current will be enhanced by any excessive clock power dissipation (an ambient temperature of ~70°C must usually be allowed for in a computer installation). A 200 x 200 element device, giving a storage capability approaching 40 000 bits will cause considerable problems in all the above limits. The chip area would be 6 mm by 3 mm even without the clock circuitry, the input/output circuitry and the regenerate/read/write circuitry. ϵ would have to be less than 2.5×10^{-4} for an $n\epsilon < 0.1$ after 400 transfers so that the binary output may be recognized without ambiguity. Finally a 10 MHz clock rate would be required in the input and output registers (this may adversely affect ϵ) in order to give a recycle time of 4×10^{-3} s. This storage or recycle time is close to the dark current limitations (section 4.3) that may be expected in a production device under practical operating conditions if there is to be no error in recognizing the binary output (Bhandarkar, 1975).

If access time becomes important in applications requiring small blocks of information (~ few hundred bits) the SPS organization is not suitable. It is then necessary to use the loop or LARAM organization but the average chip area per bit of information is significantly larger owing to the more complex peripheral circuitry and the inability to use electrode/bit clock organizations.

In all these memory organizations there is a potential to improve the storage density by using multilevel charge storage instead of the simple binary storage.

8.2.2 Clock power dissipation

When a clock electrode is raised to a potential, V_c, the ultimate charge, q_B, stored on the electrode originates in a constant voltage source equal to V_c and flows through the resistance of the driving circuitry. The energy, E_c, lost from the voltage source is

$$E_c = q_B V_c \tag{8.1}$$

If reactive driving circuitry were employed, the majority of E_c would be recovered when the clock voltage was reduced to its low level. In practice the driving circuitry is usually resistive so that equation 8.1 describes the energy dissipated by the clock for one electrode during one phase of one clock period. For an empty well with an area of 30 μm x 15 μm and V_c = 5 V, q_c will be $\sim 10^{-13}$ C for a substrate doping a little greater than 10^{20} m^{-3} (section 2.2). In this case,

$$(E_c)_{empty} = 5.10^{-13} \text{ J} \tag{8.2}$$

At a clock frequency, f_c, of 1 MHz the clock power dissipation, P, would be given by

$$(P)_{empty} = (E_c f_c)_{empty} = 0.5 \ \mu\text{W} \tag{8.3}$$

If the saturation charge were stored in the potential well, q_c could typically be ten times larger (section 2.7) so that

$$(E_c)_{full} = 5.10^{-12} \text{ J}, \tag{8.4}$$

and at 1 MHz

$$(P)_{full} = 5 \ \mu\text{W} \tag{8.5}$$

From each of the above cases, the energy dissipation per bit or the power dissipation per bit is obtained by multiplying by the appropriate factor to allow for the average number of electrodes required to store each bit.

The SPS arrangement has an advantage over, for example, the serpentine arrangement operating at the same data rate. If we consider a 16 kilobit memory operating at a 1 MHz bit rate with two electrodes per bit the maximum power dissipation for the serpentine organization is

$$(P)_{serpentine} = 0.16 \text{ W} \tag{8.6}$$

The same data rate and data storage, N_b, can occupy a SPS memory with M columns and N stores per column where

$$N_b = M + MN \tag{8.7}$$

In equation 8.7 the total data storage in the input and output registers is M units. If the clock frequency is f_c in the input and output registers the corresponding column frequency is f_c/M. Therefore the total power dissipation is

$$(P)_{SPS} = M(E_c f_c)_{input} + M(E_c f_c)_{output} + MN \, E_c f_c/M \tag{8.8}$$

If we neglect the half of the sums of power dissipation in the input and output registers which always occurs with empty wells equation 8.8 becomes

$$(P)_{SPS} \simeq (M+N) E_c f_c \tag{8.9}$$

From equations 8.7 and 8.9, $(P)_{SPS}$ is a minimum when $M = N_b^{\frac{1}{2}}$ and $N = N_b^{\frac{1}{2}} - 1$. Therefore the clock power dissipation in the optimized conditions for 1 MHz bit rate and $N_b = 16\,000$ is

$$(P)_{SPS} \simeq 2.53 \text{ mW} \tag{8.10}$$

The difference in power dissipation between equations 8.6 and 8.10 becomes important for composite storage systems with many megabits of data storage. Combining equations 8.9 and 8.7 the ratio of the power dissipation in an SPS organization compared with that in a simple serial organization is $(M + N)/[M(1 + N)]$. For large arrays with $N \gg 1$ the improvement factor is approximately $(M + N)/MN$.

Further advantages arise from a low power dissipation. Under quiescent conditions the stored data must be recirculated and refreshed at intervals set by the dark current degradation of the stored charge. The strong increase of mean dark current with temperature has already been illustrated in Fig. 7.7 where an increase of ambient temperature by 30°C above room temperature caused approximately an order of magnitude increase of dark current. In the quiescent state the memory will be refreshed at the minimum possible rate to avoid undesirable power dissipation. An increase of 30°C in the ambient temperature would then increase the necessary regeneration rate by ten times. In turn the increased power dissipation would increase the device working temperature causing a further increase of dark current.

Low operating temperature is also advantageous for an extended operating life because the mean time between failure of devices decreases rapidly with increasing temperature. Device failure from processes within the device is caused by changes of internal structure owing to diffusion or similar thermally activated processes. The mean time between failures then depends on the working temperature and the activation energy of the failure process as illustrated in the Arrhenius plot of Fig. 8.8.

8.2.3 The output binary decision

The simplest recognition of one output bit as a 'zero' or a 'one' requires a differential amplifier in which the output is compared with half the saturation charge of one well. A so-called half-charge generator consisting of a single CCD input cell with half the area of the input cell to the main CCD store may be used for this function. Unfortunately, the design of the decision circuit is not so simple. The effects of dark current must be allowed for when the storage time is long. If the dark current provides a fraction, p, of the saturation charge during the entire storage period it is necessary to make the input charge, corresponding to a 'one', no greater than $(1 - p)$ of the saturation charge to avoid charge overflow into neighbouring stores. The optimum decision point would then be $(1 + p)/2$ of a saturation charge.

The larger charge handling capacity of a surface channel CCD would appear advantageous to reduce the error in the output decision. However, the necessity for a fat zero reduces the available dynamic range and the inevitably higher charge-transfer inefficiency has a similar effect. The charge smearing caused by surface states at the channel edges (section 4.2.2) causes the optimum point of the output decision to depend on previous output bits. Following a large string of 'ones' the surface states at the channel edges will be effectively filled up and can cause a relatively large amount of charge to be smeared into the first following 'zero'. These same surface states will not contain charge if the previous output has been a large string of 'zeros'. These problems require the decision

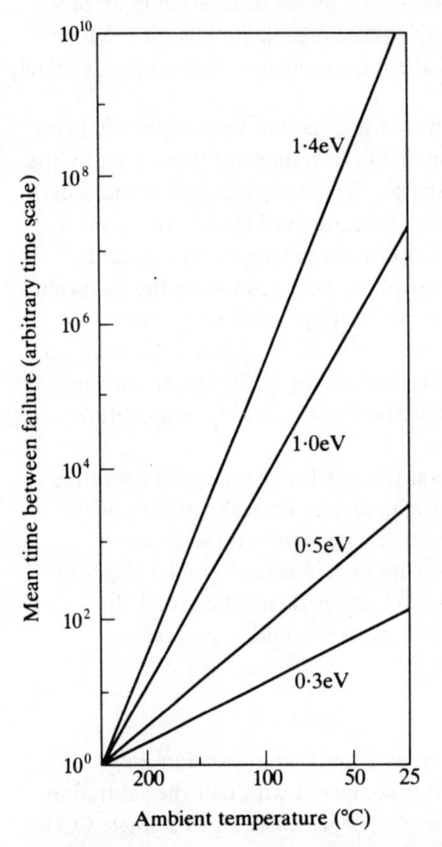

Fig. 8.8 An Arrhenius plot showing the dependence of relative failure rate on temperature for various activation energies.

point to vary in sympathy with the averaged charge output history to maintain optimum decisions. Alternatively, it may be necessary to code the data so that 'zeros' and 'ones' are always occurring with nearly equal probability to avoid the problem. Such a solution would probably involve more than one storage site being required for each bit of data. These considerations will often lead to the adoption of a buried channel CCD design where the problems are more predictable even though the saturation charge is smaller. More detailed considerations have been given by Amelio (1976).

8.3 Logic applications

The high packing density and good signal charge isolation in CCD make them attractive for logic applications. Their serial nature limits their flexibility when compared with conventional logic circuits and they have their greatest potential use in situations where parallel data streams need continuous processing. A simple way to recognize the addition of two 'ones', represented by nearly full wells, is to collect the charge overflow when both are placed in the same well. This feature is exploited in the AND/OR gate of Fig. 8.9 (Mok and Salama, 1974).

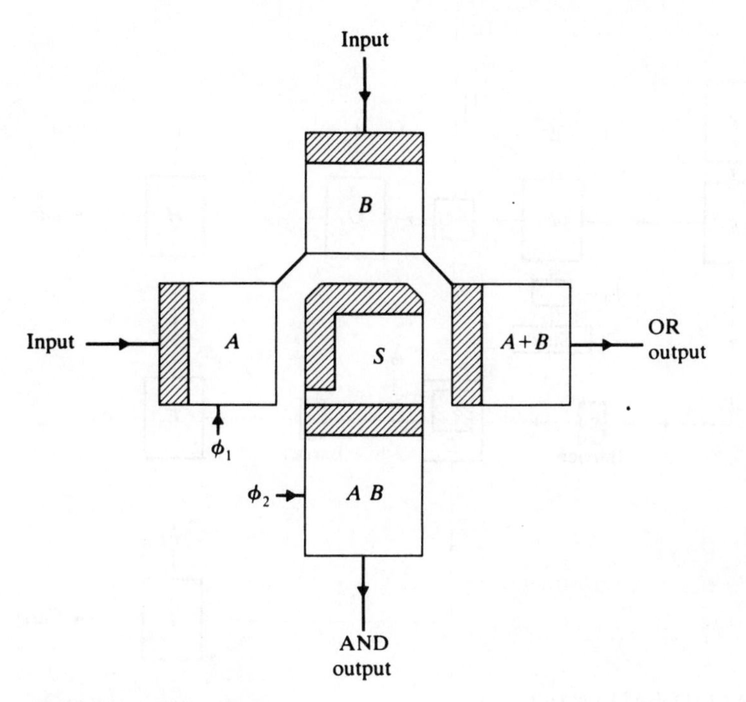

Fig. 8.9 A CCD AND/OR gate.

The five electrodes have the transfer characteristic of two-phase electrodes with the higher potential region (the part not containing minority carriers) indicated by the shading. Electrode S can cause transfer to occur into $A + B$ or into $A\,B$ depending on their relative clock potentials. For the present application A, B and $A + B$ are connected to clock phase ϕ_1, S and $A\,B$ are connected to ϕ_2. When ϕ_1 is on and ϕ_2 is off input 'zeros' or 'ones' are transferred into A and B. As ϕ_2 comes on and ϕ_1 goes off the contents of A and B are transferred into S. If both A and B contained a 'one', charge will overflow into $A\,B$ and give an AND output of 'one' at the next transfer. Also at this next transfer any of the contents of S will be transferred into $A + B$. On the next transfer from $A + B$ there will be an OR output of 'one' if there was an initial charge in A or B.

A more ambitious logic circuit is the full adder (Zimmerman and Allen, 1976) shown in Fig. 8.10. This device has to sum two binary signals, *a* and *b*, and the carry signal, *g*, from the previous less-significant-digit summation. All CCD storage locations *C, D, E, F, G, H* and *I* have the same charge capacity as *a*, *b*, and *g*. The two potential barriers allow charge to spill from *C* to *D* or from *D* to *E* if the preceding storage location is overfilled. At the start of the process the clock phase ϕ_1 is applied to *C* so that the charge in *a*, *b* and *g* enters *C*. If there are two 'ones', charge also spills over and fills *D*. If there are three 'ones' it spills

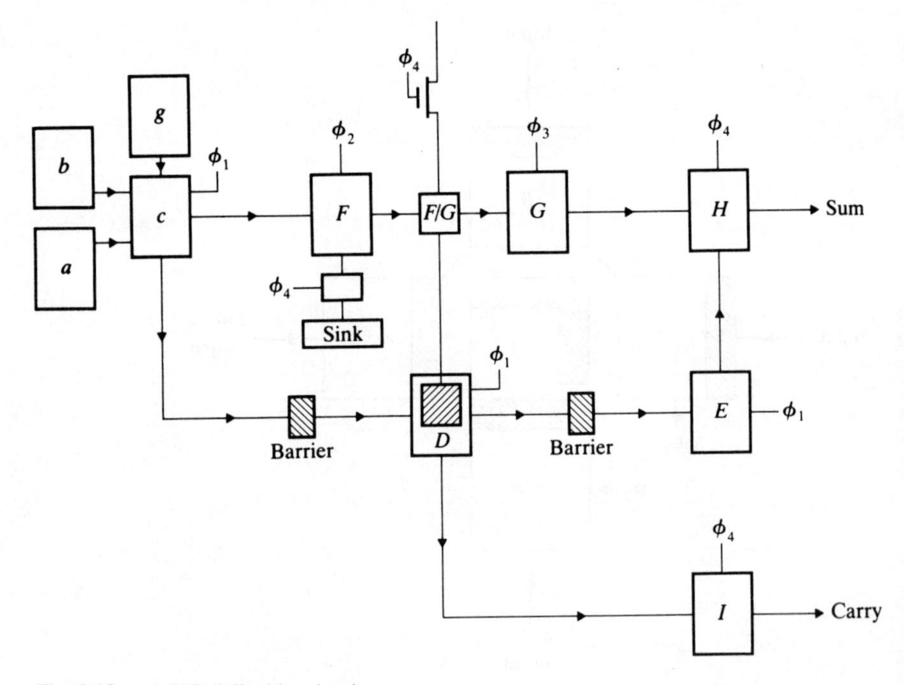

Fig. 8.10 A CCD full adder circuit.

further and fills *E*. The presence of charge in *D* changes the potential on the floating gate, *F/G*, so that charge cannot be transferred from *F* to *G* at any stage in the subsequent clocking sequence. *F/G* has to be reset before the transfer sequence begins so that charge transfer could occur from *F* to *G*, at the appropriate clock phase, providing there was no charge stored in *D*. The second clock phase, ϕ_2, allows charge to transfer from *C* to *F*. The third clock phase, ϕ_3, transfers charge from *F* to *G* providing *F/G* is open. This transfer will only occur if there is a single one in the initial input from *a*, *b* or *g*. When the fourth clock phase, ϕ_4, is applied the contents of *G* or *E* are transferred to *H*, the contents of *D* are transferred to *I* and the contents of *F* are destroyed in the sink. In this way three 'zeros' at the input give 'zero' output in both the *sum* and *carry* lines. A single 'one' in the input gives a 'one' in the *sum* output and a 'zero' in the *carry* output. Two 'ones' at the input give a 'zero' in the *sum* output and a 'one' in the *carry* output. Three 'ones' in the input give a 'one' in the *sum* output and a 'one' in the *carry* output.

The implementation of a multiplier for two binary numbers with up to four binary digits each is illustrated in Fig. 8.11. If the two numbers are $a_4\ a_3\ a_2\ a_1$

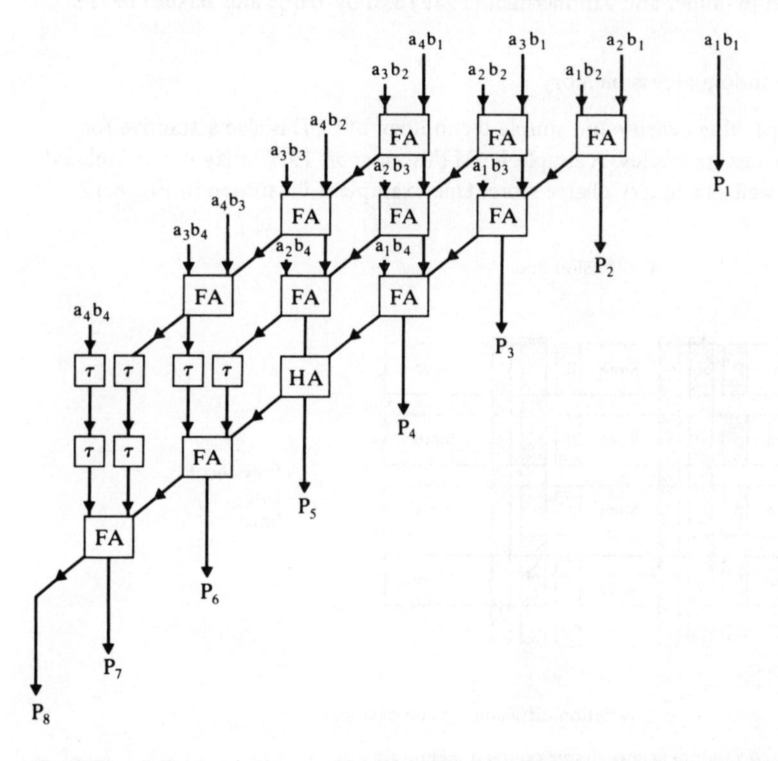

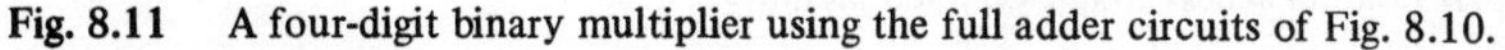

Fig. 8.11 A four-digit binary multiplier using the full adder circuits of Fig. 8.10.

and $b_4\ b_3\ b_2\ b_1$, where the a and b are 'ones' or 'zeros' the multiplication involves summations defined as follows

$$\begin{array}{cccccccc}
 & & & & a_4 & a_3 & a_2 & a_1 \\
 & & & & b_4 & b_3 & b_2 & b_1 \\
\hline
 & & & & a_4b_1 & a_3b_1 & a_2b_1 & a_1b_1 \\
 & & & a_4b_2 & a_3b_2 & a_2b_2 & a_1b_2 & \\
 & & a_4b_3 & a_3b_3 & a_2b_3 & a_1b_3 & & \\
 & a_4b_4 & a_3b_4 & a_2b_4 & a_1b_4 & & & \\
\hline
P_8 & P_7 & P_6 & P_5 & P_4 & P_3 & P_2 & P_1 \\
\hline
\end{array} \qquad 8.11$$

The column additions to give the output digits P_1, P_2 etc. must be carried out from right to left so that the carry digits from one column to the column of the next most significant digit is always available at the input to each adder. In Fig. 8.11 each column addition is carried out in successive pairs. In some cases only a half adder, HA, is required owing to the presence of only two inputs. There is also the requirement for some delay units, τ, so that successive steps in the calculation maintain their synchronism. Integration of these functions into more complicated processing units such as fast Fourier transformers has been considered by Miller and Zimmerman (1975) and by Barbe and Baker (1975).

8.4 CCD random access memory

The high packing density and simple technology of CCD is also attractive for random access memories. A simple RAM (Tasch *et al*, 1976) may use an isolated potential well as a binary charge store. One example is illustrated in Fig. 8.12

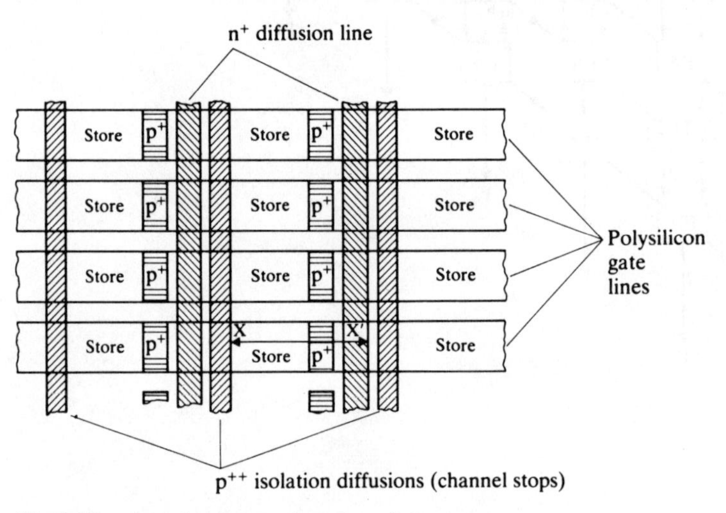

Fig. 8.12 A random access charge-coupled memory.

for a p-type substrate. Each linear array of such stores can be isolated from a linear n^+ diffusion by a perpendicular MOS gate line (the isolating potential is under the p^+ region). By making the gate region with a p^+ implant, the gate can be allowed to overlap the store region and also control its surface potential. In the absence of stored charge the surface potential energy in the store region will always be lower than that under the p^+ region just as in the case of the two-phase clock electrode structure discussed in section 2.5.1. In operation the gate line can have one of three potentials while the n^+ lines, which serve the same purpose as an input diode diffusion of a simple CCD and control the surface potential underneath, can have one of two potentials. In the store state the voltage on the gate line has its intermediate value so that the surface potential along a line XX′ in Fig. 8.12 has the form shown in Fig. 8.13. The potential under the p^+ region acts as a potential barrier to the n^+ line in either of its states. In order to write into the store, the gate line voltage is increased so that the surface potential has the form shown in the *write* condition. If the n^+ line has its lower voltage applied, charge flows into the store and a full well of charge remains after the gate line voltage is restored to its store condition. The process has strong similarities to the fill and spill input process discussed in section 5.2.3. No charge would have entered the store if the higher voltage had been applied to the n^+ line. In order

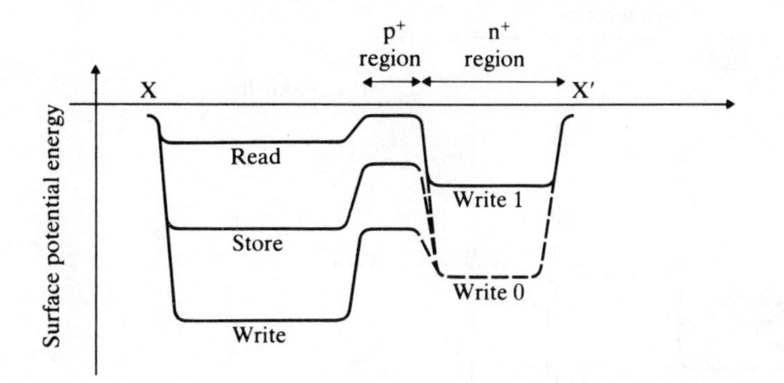

Fig. 8.13 Surface potential profiles in a single cell of Fig. 8.12.

to read the store, the gate line voltage is reduced to zero so that the surface potentials under both the p^+ and store regions collapse to a low value and are nearly equal. Charge is then transferred to the n^+ line which is acting like the output diode of a simple CCD.

The simple technology and high packing density of the above store is attractive for its use in random access memories. Its disadvantage is the three-level bias requirement for the polysilicon gate lines which makes it incompatible with existing computer stores based on the one transistor MOS RAM. Accordingly, the device whose cell construction is illustrated in Fig. 8.14 has been evolved (Tasch *et al*, 1976). It only requires two levels of bias voltage applied to the polysilicon gate to produce the two surface potential configurations in AA^1 as shown in Fig. 8.14. The neighbouring electron potential energy in the n^+ diffusion line between A^1 B can also independently take one of two values depending on its bias voltage. The cell design in AA^1 must produce a storage well isolated from the n^+ diffusion line when zero bias voltage is applied to the polysilicon gate, but application of a bias voltage on this continuous strip electrode must remove the potential wall at the right hand side of AA^1. This apparently formidable requirement is achieved in a rather clever way. As was shown in section 2.2 there is a component of the electric field discontinuity at the insulator–semiconductor interface which is caused by spurious fixed interface charge (equation 2.15). If this charge is positive the electric field difference at zero bias voltage can only exist with a positive voltage at the interface (causing a negative surface potential energy for electrons) and a corresponding voltage of

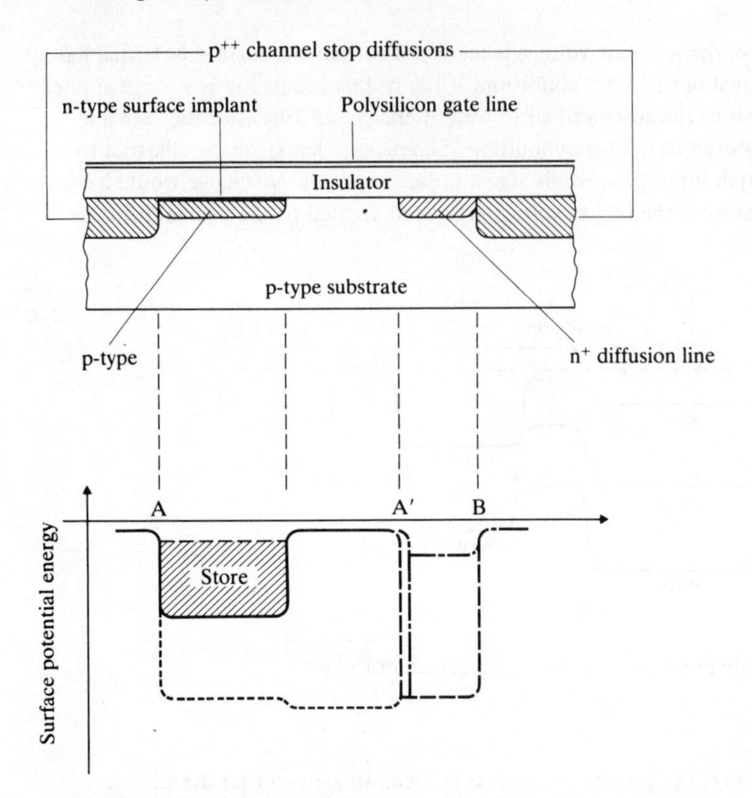

Fig. 8.14 The cross-section and surface potentials of a charge-coupled RAM cell which is compatible with existing one-transistor RAM cells.

opposite polarity across the insulator. The zero bias storage region in Fig. 8.14 is created by exploiting this effect with a thin layer of donor atoms implanted into the p-type substrate. In order that the potential wall of the storage region is removed on application of a bias voltage to the polysilicon gate, the surface potential under the non-implanted region in AA^1 must decrease more sensitively with increasing bias voltage than is the case for the storage region. This is achieved by a p-type doping of the storage region which is greater than in the rest of the substrate so that a greater proportion of the bias voltage in this region is dropped across the insulator. The same conclusion can be reached by inspection of Fig. 2.5. The n^+ diffusion line between A^1 B acts independently and like the input diode of a simple CCD as described in section 2.14.

Information is written into the store when bias voltage is applied to the polysilicon gate electrode (the word line). If a 'one' is to be written, there is no bias voltage on the n^+ diffusion line (the bit line) so that electrons flood into the storage region. The bias voltage is then removed from the polysilicon gate line and a full well of charge is stored. To write a 'zero', bias is applied to the n^+ diffusion line so that electrons cannot enter the storage region when bias is on the polysilicon gate line.

The reading process is carried out in the reverse fashion with the n^+ diffusion line acting like the destructive output diode of a conventional CCD, as described in section 2.16. This requires bias to be applied to the n^+ diffusion line. When a bias voltage is applied to the polysilicon gate line the stored charge will flow out into the n^+ diffusion line and be sensed as a 'one'. If there is no charge flow, the store had contained a 'zero'. This electrical organization of the read/write process is compatible with existing one-transistor MOS RAM organizations, so it can act as a direct replacement.

9

An Outline of Fabrication Techniques

9.1 Some components of the MOS process

The rapid progress of charge-transfer techniques owes much to the previous existence of MOS integrated-circuit technology. In turn, the new demands of CTD on MOS technology have broadened the art. The objective of this chapter is to give a brief and simple outline of those fabrication techniques relevant to the design of CCD devices described in previous chapters. The details have many similarities to those of MOS devices used for peripheral circuits or for bucket-brigade devices. Further details have been described by Wolfendale (1972) and Allison (1975).

Transfer of charge is more efficient with electrons than with holes owing to the higher mobility of electrons. In silicon MOS technology this implies that a p-type silicon substrate must be used so that minority carriers in the surface channel device are electrons. In buried channel technology the substrate must again be p-type with an n-type surface layer. A further advantage arises for a silicon dioxide insulator on a p-type silicon substrate. The fixed charge stored near the interface is positive so that the surface region of the silicon is held in depletion even at zero bias volts. On an n-type substrate this same positive charge would cause the surface to be held in accumulation so that the non-store condition of a CCD well would have to be defined with a finite bias to avoid undesirably large charge loss and surface state smearing as described in section 4.4.

The individual elements of the device are defined by photolithographic techniques in several successive operations. The partially made-up device is covered with a photographic emulsion, known as photoresist, which is spun at high speed before it dries, in order to give a uniform coverage. The desired pattern is defined by exposing the emulsion to an ultra-violet light passed through a pattern which has previously been photographically reproduced on a glass plate (Fig. 9.1). The

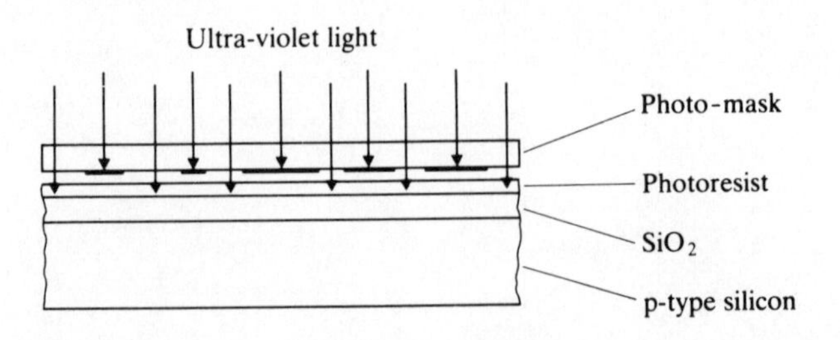

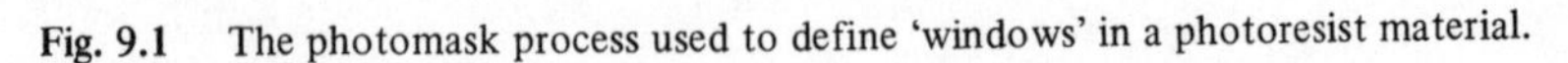
Fig. 9.1 The photomask process used to define 'windows' in a photoresist material.

ultra-violet exposure hardens the photoresist and the unexposed regions can be removed by washing in a suitable solvent. This photographic masking technique is suitable only if the subsequent processing is not at too high a temperature. For example, it is satisfactory if acid etching is to be used to remove regions of the oxide insulator or semiconductor. After the etching process, the remaining hardened photoresist is removed and, after thorough cleaning, the slice is ready for the next stage.

High-temperature masking, as in the diffusion of impurities into the semiconductor, may be carried out by the oxide itself as previously defined by a photoresist and etch procedure (referred to as photoengraving).

Two types of silicon dioxide are commonly used in the MOS process. The initial covering of the entire substrate is achieved with a 'thick' oxide. This is grown to a thickness of about 1.5 μm by placing the silicon substrate in a furnace at a temperature of, typically, 1100°C in a steam ('wet') atmosphere. The steam causes thermal oxidation of the surface to occur much more rapidly than it would in dry oxygen alone. The thickness of this oxide is made as great as possible so that the interconnections later applied to regions outside the desirable device areas do not cause parasitic turn-on (low surface potential energy) of the surface region of the underlying semiconductor. Such turn-on could cause undesirable connections between isolated regions of the device. A limit is set to the oxide thickness by the difficulty of accurately defining regions by etching through the thick oxide. Too great a thickness may lead to excessive undercutting of the oxide. Furthermore, it is difficult to deposit aluminium conductors, which may be required in subsequent processing steps, down the sides of the thick oxide. This problem is a frequent cause of device failure owing to interconnection breaks.

The second type of silicon dioxide insulator is the 'thin' oxide which is produced in dry oxygen. It has a greater freedom from fixed charge which affects the threshold voltage of the conducting channel. Typically its thickness is between 0.1 μm and 0.15 μm. It is made as thin as possible to allow control of the conducting channel of the semiconductor by as small a voltage as possible. The limitation on thickness is set by the necessity to avoid 'pinholes' arising from thickness non-uniformity which causes leakage current problems. It must also be thick enough to avoid electrical breakdown when all the bias voltage is dropped across the insulator under full well conditions of a CCD when the depletion layer of the semiconductor is non-existent.

9.2 Aluminium gate devices

The first generation of CCDs used the so-called aluminium gate technique. The initial processing stage is to define the channel for charge transfer. This can be done with a thick oxide with the channel region subsequently removed and replaced with a thin oxide. In this way the region outside the channel is never significantly depleted so that a surface potential wall exists around the channel. The disadvantage of this approach is the difficulty of depositing aluminium conductors down the channel sides at a later stage.

An alternative approach is to use a channel-stop or isolation diffusion as shown in Fig. 2.8. This is a p^+ diffusion along the sides of the channel in the p-type substrate. The p^+ diffusion stops the formation of a depletion layer of any significant width owing to its high majority carrier density so that, again, a surface potential wall exists around the channel. The masking operation can be carried out by photoengraving through the thin oxide subsequently required for insulation over the channel. The diffusion is often carried out by first depositing boron from boron trichloride, or another suitable boron compound, on to the slice in a furnace at a temperature below the oxidation temperature. The oxide acts as a mask and some boron diffuses into the exposed surface of the silicon. The remainder of the boron is removed from the surface of the slice by etching and the slice is replaced in a furnace at a higher temperature so that the boron in the surface of the silicon diffuses into the semiconductor to the desired depth, which may be 2 μm to 3 μm. Part of the diffusion occurs sideways underneath the oxide and this helps to 'seal' the semiconductor, so aiding reliability. At the same time an oxide coating forms over the diffused regions and protects it from the further processing steps. The input and output diodes are formed next, in a similar way with the appropriate dopant, by using n^+ diffusions through the appropriate photoengraved regions. The slice cross-section at this stage is illustrated in Fig. 9.2. After opening contact 'windows' to appropriate parts of the semiconductor by photoengraving, electrodes and interconnections are formed by evaporating aluminium over the entire slice to a depth of about 1 μm. The separate parts are then defined by photoengraving. Finally the whole structure is sealed by depositing an oxide insulation from the vapour phase. The cross-section now has the form shown in Fig. 9.3. Cross-over points of interconnections have to be carried out by one interconnection 'passing' beneath the other through a diffused interconnection in the semiconductor.

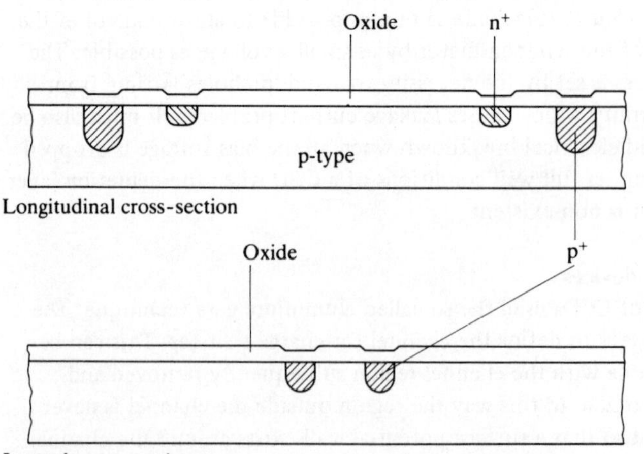

Fig. 9.2 The initial diffusion processes to define the channel stops and input/output diode regions in a CCD.

Two defects of the aluminium gate technique are device failures due to breakages in the interconnections at steep slopes in the oxide pattern and the finite separation of each electrode. The latter problem causes potential humps or hollows (sections 2.11 and 4.5). Several neat techniques have been suggested to make the gaps as small as possible without producing short circuits (Brown and Perkins, 1974; Beynon *et al*, 1974) but a better solution is provided by the polysilicon gate technology described in the next section.

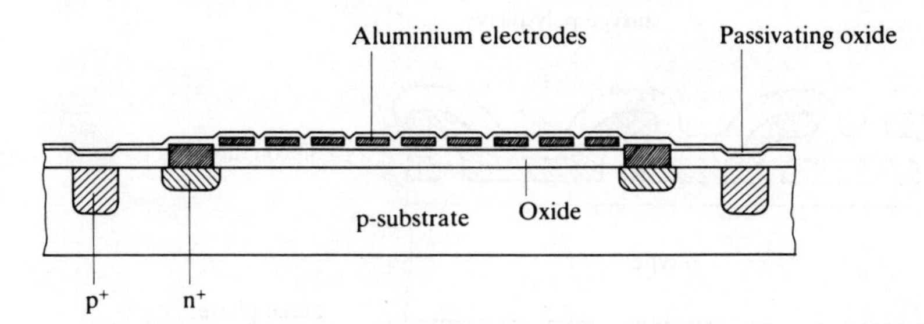

Fig. 9.3 A cross-section of an aluminium gate device.

9.3 Polysilicon gate technology

When silicon is deposited from the vapour phase on to a silicon dioxide surface it has a polycrystalline form. Just as in the crystalline form, it can be doped to give low resistivity material. Its resistivity is always much higher than that of aluminium but it is a very good conductor when compared with silicon dioxide. As an electrode and interconnection material it has some advantages over aluminium. It is easily deposited on to the sides of oxide edges without leaving fatal breaks in an interconnection. More important it can be oxidized by the normal silicon oxidation processes. This later feature allows easier realization of interconnection cross-overs and it allows successive deposition of neighbouring electrodes with overlap but without short circuit connection.

Realizations of three- and four-phase gapless structures are shown in Fig. 9.4. The processing up to the electrode deposition is much the same as it was for the aluminium gate structure. The first polysilicon deposition is carried out over the entire slice on to the previously deposited thin oxide. The first set of electrodes (marked 1 in Fig. 9.4) are then defined by removing the rest of the polysilicon by photoengraving. The slice is next given a thin oxide treatment and this is followed by the second polysilicon deposition. The second set of electrodes are cut by photoengraving and they overlap the first set of electrodes. Insulation is provided by the previous oxidation and there is a considerable reduction in the accuracy required of the mask aligning machine with consequent improvement in the yield of useful devices. Oxidation is again carried out after this second deposition. If a three-phase device is in production a third deposition of polysilicon is made and oxidized after photoengraving. The last deposition seals the device against external charge entering the oxide near the surface of the p-type silicon so that day-to-day variations of device performance do not occur. Owing to the significant resistivity of the polysilicon interconnections, even when doped, the longer interconnections away from the immediate area of the device may be made with aluminium. Reference to Fig. 9.4 indicates that the electrode gap is the thickness of the thin oxide and is approximately 0.1 μm compared with 1 μm to 2 μm for a production aluminium gate device. The consequence of this narrow

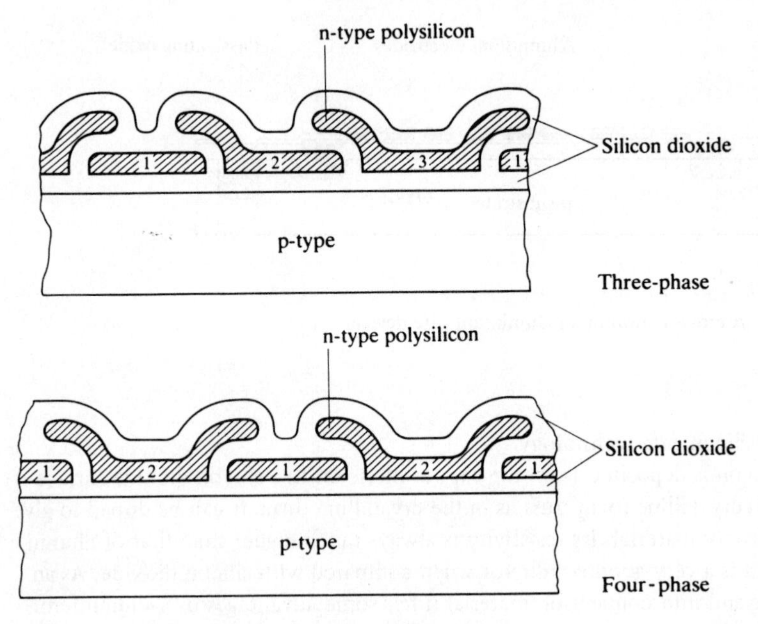

Fig. 9.4 A cross-section of two types of silicon gate device.

gap and its effective sealing from the atmosphere is a device which does not suffer from potential humps and hollows.

9.4 Two-phase structures

Two-phase devices can be constructed with four-phase electrodes with neighbouring pairs connected to the same clock waveform except for a d.c. offset to give directionality. If directionality is built into the electrodes (section 2.5) it may be achieved with a stepped oxide or a p^+ implant. The latter may be achieved in the same way as the channel stop diffusion or by ion implantation as described in the following section. In both cases appropriate masking is required. If the stepped oxide technique is used, the thicker part is grown over the entire slice and windows are photoengraved for the thinner parts. A thinner oxide is then grown in these windows and electrodes are deposited later. The step in the oxide may cause some difficulties with continuity of the deposited electrode.

9.5 Buried channel structures

The n-type region on a p-type substrate required for buried channel devices may be grown epitaxially from the vapour decomposition of silane (SiH_4) at about 1050°C. It forms a single-crystal continuation of the substrate crystal.

Alternatively, a thin region of the surface of the p-type substrate may be converted to n-type by the technique of ion implantation. The dopants are introduced by 'firing' high energy ions (typically 100 keV energy) at the substrate with an ion beam accelerator. The penetration of the ions is determined by their energy and selective masking can be carried out with an aluminium layer or a thick oxide. After implantation, an annealing process at an elevated temperature is usually required so that the ions can take their correct place in the crystal lattice.

References

Allison, J. A. 1975 *Electronic integrated circuits – their technology and design*. London: McGraw-Hill.

Amelio, G. F. 1972 Computer modelling of charge-coupled device characteristics. *BSTJ*, **51**, 705–73, March 1972.

Amelio, G. F. 1974 The impact of large CCD image sensing area arrays. *Proc. CCD '74*, 133–52, see gen. refs.

Amelio, G. F. 1976 Device design for CCD memory. *Proc. CCD '76*, 159–78, see gen. refs.

Baertsch, R. D. 1976 Charge-coupled and charge injection device performance tradeoffs. *Proc. CCD '76*, 66–74, see gen. refs.

Bailey, W. H., Buss, D. D., Hite, L. R. and Whatley, M. W. 1975 Radar video processing using the CCD chirp-Z transform. *Proc. CCD '75*, 283–90, see gen. refs.

Barbe, D. F. 1976 Charge-coupled device and charge-injection device imaging. *IEEE J. Solid-St. Circuits*, **SC-11** 109–14, Feb. 1976.

Barbe, D. F. and Baker, W. D. 1975 Signal processing devices using the charge-coupled concept. *Microelectronics J.*, **7**, 36–45, Dec. 1975.

Bell, D. A. 1960 *Electrical noise*. London: Van Nostrand.

Berger, J. L. and Coutures, J. L. 1976 Cancellation of aliasing in a CCD low-pass filter. *Proc. CCD '76*, 302–8, see gen. refs.

Berglund, C. N. 1971 Analogue performance limitations of charge-transfer device shift registers. *IEEE J. Solid-St. Circuits*, **SC-6**, 391–94, Dec. 1971.

Berglund, C. N. 1972 The bipolar transistor bucket-brigade shift register. *IEEE J. Solid-St. Circuits*, **SC-7**, 180–5, April 1972.

Berglund, C. N. and Boll, H. J. 1972 Performance limitations of the IGFET bucket-brigade shift register. *IEEE Trans. Electron Devices*, **ED-19**, 852–60, July 1972.

Berglund, C. N. and Strain, R. J. 1972 Fabrication and performance considerations of charge-transfer dynamic shift registers. *BSTJ*, **51**, 655–75, March 1972.

Beynon, J. D. E., Haken, R. A. and Baker, I. M. 1974 Charge-coupled structures with self-aligned submicron gaps. *Proc. CCD '74*, 92–9, see gen. refs.

Bhandarkar, D. P. 1975 Digital CCD memory trade-offs: a systems viewpoint. *Microelectronics J.*, **7**, 30–5, Dec. 1975.

Boonstra, L. and Sangster, F. L. J. 1972a Analogue functions fit neatly on to charge-transport chips. *Electronics*, 64–71, 28 Feb. 1972.

Boonstra, L. and Sangster, F. L. J. 1972b Progress on bucket-brigade charge-transfer devices. *IEEE Solid-St. Conf. Dig. Tech. Pap.*, **15**, 138–41 and 226–27.

Brodersen, R. W., Buss, D. D. and Tasch, A. F. 1975 Experimental characterization of transfer efficiency in charge-coupled devices. *IEEE Trans. Electron Devices*, **ED-22**, 40–6, Feb. 1975.

Brodersen, R. W. and Emmons, S. P. 1976 Noise in buried channel charge-coupled devices. *IEEE J. Solid-St. Circuits*, **SC-11**, 147–55, Feb. 1976.

Brown, D. M., Ghezzo, M. and Garfinkel, M. 1976 Transparent metal oxide electrode CID imager. *IEEE Trans. Electron Devices*, **ED-23**, 196–200, Feb. 1976.

Browne, V. A. and Perkins, K. D. 1974 Buried channel CCDs with submicron electrode spacings. *Proc. CCD '74*, 100–5, see gen. refs.

Buss, D. D., Bailey, W. H., Holmes, J. D. and Hite, L. R. 1975 Charge-transfer device transversal filters for communication systems. *Microelectronics*, 7, 46–53, Dec. 1975.

Buss, D. D., Bailey, W. H. and Tasch, A. F. Jr. 1974 Signal processing applications of charge-coupled devices. *Proc. CCD '74*, 179–97, see gen. refs.

Buss, D. D., Brodersen, W., Hewes, C. R., and de Wit, M. 1976 Spectral analysis using CCD. *Proc. CCD '76*, 208–18, see gen. refs.

Buss, D. D. Veerkant, R. L., Brodersen, R. W. and Hewes, C. R. 1975 Comparison between the CCD CZT and the digital FFT. *Proc. CCD '75*, 283–90, see gen. refs.

Butler, W. J., Puckette, C. McD. and Smith, D. A. 1973 Differential mode of operation for bucket-brigade devices. *Electronics Letts.*, 9, 106–7, 8 March 1973.

Campana, S. B. and Barbe, D. F. 1974 Trade-offs between aliasing and MTF. *Proc. CCD '74*, 168–76, see gen. refs.

Carnes, J. E. and Kosonocky, W. F. 1972a Noise sources in charge-coupled devices. *RCA Rev.*, **33**, 327–43, June 1972.

Carnes, J. E. and Kosonocky, W. F. 1972b Fast-interface-state losses in charge-coupled devices. *Appl. Phys. Letts.*, **20**, 261–3, 1 April 1972.

Carnes, J. E., Kosonocky, W. F. and Levine, P. A. 1973 Experimental measurements of noise in charge-coupled devices. *RCA Rev.*, **34**, 553–65, Dec. 1973.

Carnes, J. E., Kosonocky, W. F. and Ramberg, E. G. 1971 Drift-aiding fringing fields in charge-coupled devices. *IEEE J. Solid-St. Circuits*, **SC-6**, 322–6, Oct. 1971.

Carnes, J. E., Kosonocky, W. F. and Ramberg, E. G. 1972 Free charge-transfer in charge-coupled devices. *IEEE Trans. Electron Devices*, **ED-19**, 798–809, June 1972.

Chou, S. 1976 Design of a 16384-bit serial charge-coupled memory device. *IEEE J. Solid-St. Circuits*, **SC-11**, 10–18, Feb. 1976.

Chowaniec, A. and Hobson, G. S. 1975 Phase errors in transversal and recursive filters realized with charge-transfer devices. *Electronics Letts.*, **11**, 467 only, 18 Sept. 1975.

Chowaniec, A. and Hobson, G. S. 1976 A wide-band quadrature phasing system using charge-transfer devices. *Solid-St. Electronics*, **19**, 201–7, March 1976.

Chowaniec, A. and Hobson, G. S. 1977 A time-domain analysis of CCD video integrators. *Solid-St. Electronics*, **20**.

Collet, M. G. and Vliegenthart, A. C. 1974 Calculations on potential and charge distributions in the peristaltic charge-coupled device, *Philips Res. Reps.*, **29**, 25–44.

Cook, C. E. and Bernfeld, M. 1967 *Radar signals*, Ch. 1, New York: Academic Press.

Crowell, M. H. and Labuda, E. F. 1969 The silicon diode array camera tube. *BSTJ*, **48**, 1481–528, May–June 1969.

Dyck, R. H. and Jack, M. D. 1974 Low-light-level performance of a charge-coupled area imaging device. *Proc. CCD '74*, 154–61, see gen. refs.

Elliot, C. T. 1976 Infra-red CCD systems. *Proc. CCD '76* 127–44, see gen. refs.

Emmons, S. P., Buss, D. D., Brodersen, R. W. and Hewes, C. R. 1975 Anti-aliasing characteristics of the floating diffusion input. *Proc. CCD '75*, 361–8, see gen. refs.

Esser, L. J. M. 1972 Peristaltic charge-coupled devices: a new type of charge-transfer device. *Electronics Letts,* **8**, 620 only.

Fawcett, W. and Vanstone, G. J. 1976 Two-dimensional simulation of the charge-transfer process in charge-coupled devices using a particle model. *J. Phys. D.: Appl. Phys.*, **9**, 773–83.

Frey, W. 1973 Bucket-brigade devices with improved charge-transfer, *Electronics Letts*, **9**, 588–9, 13 Dec. 1973.

Goetzberger, A. 1974 Interface states in Si–SiO_2 interfaces. *Proc. CCD '74*, 47–58, see gen. refs.

Gold, B. and Radar, C. M. 1969 *Digital processing of signals*. New York: McGraw-Hill.

Holeman, B. R. and Gardner, P. 1976 Applications of CCD imagers at low light levels. *Proc. CCD '76*, 98–116, see gen. refs.

Joyce, W. B. and Bertram, W. J. 1971 Linearized dispersion relation and Green's function for discrete charge-transfer devices with incomplete transfer. *BSTJ*, **50**, 1741–59.

Kim, C. K. and Dyck, R. H. 1973 Low-light-level imaging with buried channel charge-coupled devices. *Proc. IEEE*, **61**, 1146–7, Aug. 1973.

Kim, C. K. and Lenzlinger, M. 1971 Charge transfer in charge-coupled devices. *J. Appl.*

Phys., **42**, 3586–94, Aug. 1971.

Kim, C. K. and Snow, E. H. 1972 p-channel CCDs with resistive gate structures. *Appl. Phys. Letts*, **20**, 514–15, June 1972.

Kosonocky, W. F. and Carnes, J. E. 1975 Basic concepts of charge-coupled devices. *RCA Rev.*, **36**, 566–93, Sept. 1975.

Krambeck, R. H., Walden, R. H., and Pickar, K. A. 1972 A doped surface two-phase CCD, *BSTJ*, **51**, 1849–66, Oct. 1972.

Lagnado, I. and Whitehouse, H. J. 1974 Signal processing image sensor using charge-coupled devices. *Proc. CCD '74*, 198–205, see gen. refs.

Lamb, D. R., Singh, M. P., Brotherton, S. D., and Roberts, P. C. T. 1974 Influence of surface states on the performance of three-phase charge-coupled devices. *Proc. CCD '74*, 59–66, see gen. refs.

Lampe, D. R., White, M. H., Fagan, J. L. and Mims, J. H. 1974 An electrically-reprogrammable analogue transversal filter. *1974 Int. Solid-St. Circuits Conf. Dig., IEEE New York*, 156–7, Feb. 1974.

McKelvey, J. P. 1966 *Solid-St. and semiconductor physics*. New York: Harper and Row.

McKenna, J. and Schryer, N. L. 1973 The potential in a charge-coupled device with no mobile minority carriers and zero plate separation. *BSTJ*, **52**, 669–96, May–June 1973.

MacLennan, D. J., Mavor, J. and Vanstone, G. F. 1975 Technique for realizing transversal filters using charge-coupled devices. *Proc. IEE*, **122**, 615–19, June 1975.

Majer, G. J. 1975 The Chirp-Z Transform – A CCD implementation. *RCA Rev.*, **36**, 759–73, Dec. 1975.

Melen, R. D., Shott, J. D., Walker, J. T. and Meindl, J. D. 1975 CCD dynamically focused lenses for ultrasonic imaging systems. *Proc. CCD '75*, 165–71, see gen. refs.

Michon, G. J., Burke, H. K., Brown D. M. and Ghezzo, M. 1975 CID imaging – present status and opportunities. *Proc. CCD '75*, 93–9, see gen. refs.

Miller, C. S. and Zimmerman, T. A. 1975 Applying the concept of a digital charge-coupled device arithmetic unit. *Proc. CCD '75*, 199–207, see gen. refs.

Mohsen, A. M. and Tompsett, M. F. 1974 The effects of bulk traps on the performance of bulk channel charge-coupled devices. *Proc. CCD '74*, 67–74, see gen. refs.

Mohsen, A. M., Tompsett, M. F., Fuls, E. N. and Zimany, E. J. 1976 A 16-K bit block addressed charge-coupled memory device. *IEEE J. Solid-St. Circuits*, **SC-11**, 40–8, Feb. 1976.

Mok, T. D. and Salama, C. A. T. 1974 A charge-transfer device logic cell. *Solid-St. Electronics*, **17**, 1147–54, Nov. 1974.

Rader, C. M. and Gold, B. 1967 Digital filter design technique in the frequency domain. *Proc. IEEE*, **55**, 149–71, Feb. 1967.

Roberts, J. B. G., Eames, R. and Roche, K. A. 1973 Moving-target-indicator recursive radar filter using bucket-brigade circuits. *Electronics Letts*, **9**, 89–90, 22 Feb. 1973.

Rosenbaum, S. D., Chan, C. H., Caves, J. T., Poon, S. C. and Wallace, R. W. 1976 A 16 384-bit high-density CCD memory. *IEEE J. Solid-St. Circuits*, **SC-11**, 33–40, Feb. 1976.

Sangster, F. L. J. 1970a The 'bucket-brigade delay line', a shift register for analogue signals. *Philips tech. Rev.*, **31**, No. 4, 97–116.

Sangster, F. L. J. 1970b Integrated MOS and bipolar analogue delay lines using bucket-brigade capacitor storage. *Dig. Tech. Pap. 1970 IEEE Int. Solid-St. Circuits Conf. Philadelphia*, 74 only, Feb. 1970.

Sangster, F. L. J. and Teer, K. 1969 Bucket-brigade electronics – new possibilities for delay, time-axis conversion, and scanning. *IEEE J. Solid-St. Circuits*, **SC-4**, 131–6, June 1969.

Sequin, C. H. and Mohsen, A. M. 1975 Linearity of electrical charge injection into charge-coupled devices. *IEEE J. Solid-St. Circuits*, **SC-10**, 81–92, April 1975.

Sequin, C. H., Shankoff, T. A. and Sealer, D. A. 1974 Measurements on a charge-coupled area image sensor with blooming suppression. *IEEE Trans. Electron Devices*, **ED-21**, 331–41, June 1974.

Sequin, C. H. and Tompsett, M. F. 1975 Charge-transfer devices. Suppl. 8 to *Advances in electronics and electron physics*, pp. 142–200. New York and London: Academic Press.

Shannon, C. 1949 Communication in the presence of noise. *Proc. IRE*, **37**, 10–21, Jan. 1949.

Simpson, P. I. 1975 CCD transversal filter with fixed-weighting coefficients. *Microelectronics*, 7, 54–9, Dec. 1975.

Skolnik, M. I. 1962 *Introduction to radar systems*. New York: McGraw-Hill.

Smith, D. A., Puckette, C. M. and Butler, W. J. 1972 Active bandpass filtering with bucket-brigade delay lines. *IEEE J. Solid-St. Circuits*, **SC-7**, 421–5, Oct. 1972.

Sze, S. M. 1969, *Physics of semiconductor devices*. New York: Wiley.

Tasch, A. F., Frye, R. C. and Horng-Sen Fu 1976 The charge-coupled RAM cell concept. *IEEE J. Solid-St. Circuits*, **SC-11**, 58–63, Feb. 1976.

Tchon, W. E., Elmer, B. R., Denloer, A. J., Negishi, S., Hirabaiyashi, K., Nojima, I. and Kohyama, S. 1976. 4096-bit serial decoded multiphase serial–parallel–serial CCD memory. *IEEE J. Solid-St. Circuits*, **SC-11**, 25–33, Feb. 1976.

Terman, L. M. and Heller, L. G. 1976 Overview of CCD memory. *IEEE J. Solid-St. Circuits*, **SC-11**, 4–10, Feb. 1976.

Theunissen, M. J. J. and Esser, L. J. M. 1974 PCCD technology and performance. *Proc. CCD '74*, 106–11, see gen. refs.

Thornber, K. K. 1971 Incomplete charge-transfer in IGFET bucket-brigade shift registers. *IEEE Trans. Electron Devices*, **ED-18**, 941–50, Oct. 1971.

Thornber, K. K. 1974 Theory of noise in charge-transfer devices. *BSTJ*, **53**, 1211–61, Sept. 1974.

Tompsett, M. F., Bertram, W., Sealer, D. A. and Sequin, C. H. 1973 Charge-coupling improves its image, challenging video camera tubes. *Electronics*, **46**, No. 2, 162–8.

Tozer, R. C. and Hobson, G. S. 1976a Reduction of high-level non-linear smearing in CCDs. *Electronics Letts*, **12**, 355–6, 8 July 1976.

Tozer, R. C., and Hobson, G. S. 1976b A capacitively-metered input circuit for linear operation of CCDs. *Proc. CCD '76*, 309–14, see gen. refs.

Van Valkenburg, M. E. 1955 *Network analysis*. Englewood Cliffs, N.J.: Prentice-Hall.

Wardrop, B. and Bull, E. 1977 A discrete Fourier transform processor using charge-coupled devices. *Marconi Rev.*, **40**, 1–41, First Quarter 1977.

Weimer, P. K. 1975 Image sensors for solid-state cameras. *Adv. Electronics Electron Phys.*, **37**, 181–262.

Wen, D. D. 1975 Low-light-level performance of CCD image sensors. *Proc. CCD '75*, 109–11, see gen. refs.

White, M. H. and Lampe, D. R. 1975 Charge-coupled device (CCD) analogue signal processing. *Proc. CCD '75*, 189–97, see gen. refs.

White, M. H., Lampe, D. R., Blaha, F. C. and Mack, I. A. 1974 Characterization of surface channel CCD image arrays at low light levels. *IEEE J. Solid-St. Circuits*, **SC-9**, 1–13, Feb. 1974.

White, W. D. and Ruvin, A. E. 1957 Recent advances in the synthesis of comb filters. *IRE Natn. Conv. Rec.*, part 2, 186–99.

Wolfendale, E. 1972 *MOS integrated circuit design*. London: Butterworth.

Zimmerman, T. A. and Allen, R. A. 1976 Charge-coupled device digital arithmetic functions: experimental results. *Proc. CCD '76*, 190–6, see gen. refs.

General References

1. *Proc. CCD '74*, Univ. Edinburgh, Centre for Ind. Consultancy and Liaison.
2. *Proc. CCD '75*, New York and Oxford, Learned Information.
3. *Proc. of CCD '76*, Univ. Edinburgh, Centre for Ind. Consultancy and Liaison.
4. Special issue on CCD technology, *Microelectronics J.*, 7, Dec. 1975.
5. Special issue on charge-transfer devices, *IEEE J. Solid-St. Circuits*, **SC-11**, Feb. 1976.
6. *The impact of new technologies in signal processing*, IEE Conf. Publ. 144, London, Sept. 1976.

Index